A Personal Flyer's Guide to More Enjoyable Flying

Capt. David C. Koch

The Aerospace Trust Press
www.theaerospacetrust.org

Copyright © 2013 Capt. David C. Koch

All rights reserved. No part of this book may be reproduced (except for inclusion in reviews), disseminated or utilized in any form or by any means, electronic or mechanical, including photocopying, recording or in any information storage or retrieval system, or the Internet/World Wide Web, without permission in writing from the author or publisher.

Printed in the United States of America

First Edition printing: August 2013.

Published by The Aerospace Trust Press
1400 E. Main Street
Ste. 103-195
St. Charles, IL 60174
info@theaerospacetrust.org

Cover picture: The author in the cockpit of his Stearman biplane

Cover design by Tyler Koch

ISBN-13: 978-0-9726991-0-5

*This book is dedicated to the excellent airmen
who revealed to me what airmanship excellence is,
and how to successfully pursue it.*

Capt. David C. Koch

Reviews of *A Personal Flyer's Guide To Better Flying*

For over fifty years Captain David C. Koch has dedicated his life to improving the field of aviation. "A Personal Flyer's Guide to More Enjoyable Flying" delivers solutions for personal flying's basic problems, and it provides a roadmap to improve safety and proficiency in aviation. According to Capt. Koch, "airmen who practice Airmanship 2.0 are strongly motivated to "fly right".

In "A Personal Flyer's Guide to More Enjoyable Flying", Capt. Koch also reveals how continuous airmanship development is yoked together with the knowledge and skills pilots need in today's aviation environment. Training and practice is provided to meet the needs of each pilot based on the type of personal flying he or she is doing.

The Greek myth, Icarus and Daedalus, tells the story of a young boy who flies above the island of Crete with wings made of wax. He forgets everything in the world but joy. As Icarus flies close to the sun, the heat melts the wax from his wings. Captain Koch believes that knowing your flight environment is key to situational awareness and safety. "If you can feel confident that your flying is as safe as possible, you will enjoy the experience."

Capt. Koch believes that the practice of Airmanship 2.0 leads to airmanship excellence. "A Personal Flyer's Guide to More Enjoyable Flying" is his comprehensive guide for personal flyers who want to become an excellent airman or airwoman. If the ideas in this book are adopted by a significant number of personal flyers, I believe we will see an incredibly positive change in the culture of the personal flying community.

Susan M. Glancey
New Private Pilot (March 2012)
110 hours

Some might think that "A Personal Flyer's Guide To More Enjoyable Flying" is only for experienced pilots, but as a student pilot I found it very informative. Had I read it before I started my training, I would have done many things differently. When choosing a flight school, I selected one close to my home and did not do any research on how it approached airmanship training. As a result, I spent six months flying an airplane, but it did not teach me to be an excellent airman. Learning how to fly does not end when you get your Private Pilot Certificate. You must continue your pursuit of airmanship excellence with every flight you take.

Volker Kotscha
Student Pilot

Table of Contents

Preface: *Reader's Preflight Briefing* ... 1

Chapter 1: *My Pursuit Of More Enjoyable Personal Flying* ... 9
   1950s ... 10
   1960s ... 14
   1970s ... 29
   1980s ... 53
   1990s ... 67
   2000s ... 81
   2010s ... 94

Chapter 2: *Airmanship 2.0* ... 105
   Why We Pursue Airmanship Excellence ... 105
   Airmanship Excellence Defined ... 109
   The History of Airmanship 2.0 ... 112
   Airmanship 2.0 For Personal Flyers ... 122
   Airmanship 2.0 Culture ... 140

Chapter 3: *Airmanship 2.0 In Practice* ... 151
   The Future: Palwaukee Aero Club ... 154
   Briefing 1: Palwaukee Aero Club History ... 157
   Briefing 2: Palwaukee Aero Club Structure ... 160
   Briefing 3: Palwaukee Aero Club Fleet ... 163
   The Tour: OpsCenter ... 173
   The Tour: Training Center ... 177
   The Tour: Business Offices ... 178
   The Tour: Hangars ... 179
   Briefing 4: Club Training Programs ... 185
   Briefing 5: Member Support ... 205
   Briefing 6: Club Membership ... 222
   Working Lunch At The Club ... 229
   The Tour: Fitness Center ... 233
   Q&A Session ... 238
   Back to Reality ... 248

Chapter 4: *Your Pursuit of More Enjoyable Personal Flying* ... 249

Acknowledgments ... 255

Resources ... 257

Preface: *Reader's Preflight Briefing*

> *Anyone can do the job when things are going right. In this business we play for keeps.*
>
> —*Ernest K. Gann*

Capt. Ernie Gann is one of my airmanship heroes. He started flying for American Airlines in the late 1930s. During World War II, he transferred to the U.S. Army Air Force's Air Transport Command where he flew **Douglas DC-3s**, **Douglas DC-4s** and **Consolidated C-87 Liberator Express** transports. After the war, he began a career as a novelist and screenwriter. **Gann's many books** include "The High And The Mighty", "Island In The Sky", "Fate Is The Hunter", "Blaze of Noon" and "A Hostage To Fortune".

I was very fortunate to have discovered his works early in my aviation career. As I recall, I started reading them while still in high school. And I've reread them several times since. They graphically portray not only what airmanship excellence truly is, but also how the standards that define it have evolved over time.

I've read them all and I highly recommend that you do too if you're interested in getting more enjoyment from your personal flying. As you will discover, the pursuit of airmanship excellence is the best path that I know of to a more enjoyable flying experience. It works for me and it is working for many other airmen and airwomen. It is my hope that by the time you're done reading this book, you will have decided to join us.

I chose to use the above Gann quote to open this book because it pointedly sums up the primary reason I wrote it. That purpose is to open the eye's of every personal flyer, and everyone who is considering becoming one, to the reality that if you're going to fly, you must do it right. There is no other reasonable alternative if you value your life and the lives of those who fly with you.

Throughout this book, I use the terms "personal flyer", "personal flyers" and "personal flying". So, I'd better define what I mean by "personal flyer". As far as I know, the term is not included in the Federal Aviation Regulations or the Aeronautical Information Manual. It is a term that has been around for a long time, but it is not in general use within the general-aviation community. I'm employing it in this book to precisely define the type of aviator I wrote this book for.

The aviation industry is subdivided into three categories: airline, military and general aviation. General aviation includes all the different types of flying except airline and military flying (i.e., bush flying, pipeline patrol, corporate operations, flight training). Personal flying is generally included in the general-aviation category. For the purposes of this book, I define a personal flyer as a person who has learned to fly and then pilots a personal airplane for his/her business-and-personal-mobility needs and for his/her enjoyment.

Many personal flyers enjoy the freedom and lifestyle that come from piloting an airplane, whether it is to distant locations or just around the local airport. Unlike professional pilots, personal flyers commit a sizeable portion of their fortunes and their discretionary time to their flying. They do that because they love to fly and the rewards they receive from pursuing that love are worth far more than what it costs them. In other words, the personal-flying value proposition works for them.

Unfortunately, we are experiencing a rapid decline in the population of our personal-flying community here in the U.S. This is due to many factors that we'll explore in this book, but it all boils down to the fact that the current personal-flying value proposition is becoming increasingly less attractive for current personal flyers and those who would like to join our community if they could only find a personal-flying value proposition that works for them. I believe that increasing the enjoyment personal flyers get from their flying is the key to making that value proposition more attractive to them. This book puts forward a new model for personal flying that is designed to significantly increase the enjoyment of personal flying.

That new model includes certain airmanship beliefs, values, practices, policies and procedures that can make personal flying much more enjoyable. And one of the reasons it is more enjoyable is that it is much safer. It's difficult to enjoy flying if you know that the flying you're doing is not as safe as it could be. For far too long now, the personal-flying community has prepared pilots to essentially "drive" an airplane. And as it turns out, almost anyone can learn to do that. However, the historical and current personal-flying accident rates show conclusively that many personal flyers do not have the airmanship skills and knowledge they need to handle things when they aren't going right.

Every day in the U.S. at least one person on average loses his or her life in a general-aviation accident. In 2011, 485 people died in mostly personal-flying accidents. That's an average of 1.3 per day. I estimate that the number of potentially fatal near-accidents is at least ten times higher. This record clearly shows that in personal flying, we play for keeps. Unfortunately, a significant number of personal flyers simply don't now how to play the game properly.

The pilots involved in these fatal accidents were mostly successful professionals (in non-aviation industries) who were perfectly capable of learning how to "fly right". Unfortunately, they all were practicing an old and outdated form of airmanship that I call Airmanship 1.0. Regrettably, their naive and unsuspecting passengers paid the ultimate price for trusting them. These pilots were not really fully qualified to confront the airmanship challenges they were faced with.

These accidents weren't caused by mysterious unknown forces. Rather, they resulted from insufficient airmanship capability in the face of hazards, that when confronted, presented the pilot with airmanship challenges that he or she was just not up to handling. And the really tragic aspect of this dismal picture is that every one of the causes that contributed to these accidents has been known to the personal-flying community for many decades, along with the means to avoid them.

It is my belief that what I describe in this book as Airmanship 2.0 will reduce the personal-flying accident rate to as close to zero as humanly possible for the personal flyers who opt to practice it. This may sound like an unrealistic hope. However, the major U.S. airlines have reduced their fatal accident rate to nearly zero primarily by implementing the practice of Airmanship 2.0 in their flight operations. My research for this book indicates that if the personal-flying community can reign in these avoidable accidents, everyone's enjoyment of the personal-flying experience will be significantly enhanced.

I also wrote this book to generate enthusiasm for a strategy for ensuring the continued viability of personal flying in the U.S. I've been a personal flyer since I started flying in 1958. Even during my years as an airline, military and corporate pilot. As I will explain in more detail later in this book, I became very concerned about the future of personal flying about four years ago, and since then I've been working on finding viable solutions to the basic problems that are rapidly sucking the life force out of it.

For example, in my opinion, the greatest threat to the future of personal flying in the U.S. is the rapidly dwindling population of the personal-flying community. To really get a feel for how critical this challenge is, we need to look back at the numerical high point of personal flying that took place in the early 1980s. At that time, there were over 800,000 FAA-certificated pilots in the U.S. I estimate that over 700,000 of them were personal flyers. There are no really accurate numbers published by the FAA for what I define as a personal flyer. So, I'll use the fuzzy numbers we do have to show the order of magnitude of the problem. These personal flyers each flew around forty hours per year on average. Today, it is estimated that there are less than 200,000 active personal flyers in the U.S. And we're flying on average only about twenty hours a year—roughly half as much.

This rapid reduction in the personal-flying customer base has driven up the cost of flying at a rate that significantly exceeds the rate of inflation. My estimate is that most of the cost factors related to personal flying (i.e., fuel, insurance, hangar, aircraft maintenance, flight training and aircraft rental) have increased by about 30% over the rate of inflation. As we all learned in Econ 101, as the customer base for a product or service decreases, the unit price of the product or service increases.

However, the biggest cost factor today in personal flying is the cost of the airplane you're flying. A major cost factor in the price of a new personal airplane is the number of aircraft produced by the manufacturer of the aircraft. As a given manufacturer produces more aircraft, the overhead burden (especially product liability premiums and certification costs) assigned to each new aircraft produced by that manufacturer is reduced. In the mid-1980s, aircraft manufacturers Cessna, Beech and Piper combined were churning out over 25,000 single-engine-piston personal aircraft a year. Today, all of the manufacturers of personal aircraft combined are selling less than 1,000 single-engine-piston personal aircraft a year.

The result of this drastic reduction in personal-aircraft production is revealed in the cost of a new personal airplane. In 1985, you could purchase a new top-of-the-line single-engine personal aircraft (i.e., a **Cessna 182 Skylane**) for around $76,000. If we apply a typical CPI (Consumer Price Index) inflation factor to the 1985 price, today's price for a new Skylane would be around $164,000. However, the list price on a new 2013 Skylane is around $400,000. In other words, a new Skylane now costs $236,000 more than what it would if the price had only increased with inflation from 1985 to 2013. Today's actual purchase price reflects an almost 70% increase over the rate of inflation.

Given the fact that cost is cited as the major barrier to personal flying by current and potential personal flyers, it is obvious that something has to be done about this situation if we want to grow our personal-flying community. As the personal-flying customer base continues to shrink, costs will continue rise and even fewer people will join our personal-flying community, and even more will drop out because of escalating costs. The airmanship term for this business trend is "graveyard spiral".

A graveyard spiral is almost always fatal unless the proper technique is applied to arrest it and begin a climb to safety. In this book, I will present to you the technique that I believe can be used to escape from this graveyard spiral. That technique is the adoption of Airmanship 2.0 by a significant number of personal flyers.

It is my belief that Airmanship 2.0 will make personal flying more affordable, more accessible and more enjoyable. It is logical to think that this will result in a rapidly growing number of personal flyers. Especially given the facts that there are now about *one hundred million* more people living in the U.S. than there were in the mid-1980s, and the continuously deteriorating value propositions of airline and road travel.

As you will see, I'm optimistic about our recovery from the graveyard spiral personal flying is currently in. That's because others and I have in the past performed the necessary elements of the recovery maneuver that is described in this book. I invite you to think of *A Personal Flyer's Guide to More Enjoyable Flying* as the manual on how to bring all of these elements together to perform the maneuver. With knowledge of the elements of the maneuver and the manual on how to perform it in hand, I have true confidence (I'll define this later in the book.) in our ability to perform the maneuver successfully. And when we complete it, the personal-flying community will be on a trajectory that will take it to new heights of personal-flying enjoyment.

Let's begin our journey through the story of how we can all enjoy personal flying more with an overview of our flightplan. It looks like this:

In *Chapter 1: My Pursuit of More Enjoyable Personal Flying*, I will tell you the story of my pursuit of it. That pursuit has taken place over the last fifty-plus years. As you will see, I believe that the pursuit of airmanship excellence is the key to personal-flying enjoyment.

In *Chapter 2: Airmanship 2.0*, we'll look at why we should practice Airmanship 2.0 or quit flying. As you'll see, many other very experienced airmen and I believe that if you're going to fly, you should do it right. In this chapter, we are also going to see if we can achieve alignment on a definition of what airmanship excellence is. We'll then use that definition throughout the remainder of this book. We'll also look at the differences between Airmanship 1.0 and Airmanship 2.0, and we'll delve extensively into what Airmanship 2.0 is. This chapter also takes a look at cultures in general and an Airmanship 2.0 culture in particular. And we'll go into what it takes to make a personal-flying culture into an Airmanship 2.0 culture.

Chapter 3: Airmanship 2.0 In Practice is a fictional story of how Airmanship 2.0 can work in the real world of personal flying. Although the story is imaginary, it is based on fact and a serious analysis of how we can implement the ideas presented in this book.

And finally, in *Chapter 4: Your Pursuit of More Enjoyable Personal Flying* we'll recap our discussion of the pursuit of airmanship excellence and look at some tips and ideas that will prove useful to you should you decide to pursue airmanship excellence through the practice of Airmanship 2.0.

This book will provide you with the information you need to make that decision. As you will see, there are many benefits that accrue to any aviator, no matter how inexperienced or senior, who faithfully adheres to the Airmanship 2.0 doctrine. My hope is that when you finish this book, you'll be convinced that there are really only two good choices that a responsible aviator can make when it comes to flying airplanes. You can either pursue airmanship excellence through the practice of Airmanship 2.0, or you can quit flying. Gen. Chuck Yeager (the first man to break the sound barrier) has been quoted as saying, "If you're going to fly, do it right." I couldn't agree more with the general's statement. If you're going to challenge the gods by defying gravity, the only logical approach is doing it right.

This book was written primarily for the people who I believe need to read it most. These readers fall into one of the following categories:

1. Non-pilots who are considering learning to fly.

2. Current pilots who want more enjoyment from their personal flying.

3. The management and staff of Airmanship Development Support Organizations.

Some additional briefing notes:

- Text in **BOLD** typeface indicates that there is a Web page that you can reference for more details. You will find the Web addresses in the *Resources* section at the back of the book. You will also find the address of a Web page on the Center For Airmanship Excellence Website that you can bookmark. The page contains hot-links to all of the resources. The Web addresses in this book were current at the time of publication. However, given the nature of the Internet, I can't guarantee that they will take you to the intended Web pages.

- Throughout the book, I use the terms "airman", "airmen", "he", "him" and "his" in a non-gender-specific manner.

- The manufacture of this book is environmentally friendly. The reduction in white space cuts down significantly on the number of pages. Also, this book is printed only on demand. That means that there are no unsold copies that will join the over 50% of all printed books that go to the shredder each year.

Well, that completes your preflight briefing. Welcome aboard and enjoy the flight.

Chapter 1: *My Pursuit of More Enjoyable Personal Flying*

> *Flying is like sex—I've never had all I wanted but occasionally I've had all I could stand.*
>
> —**Stephen Coonts**, *The Cannibal Queen*

I believe that the pursuit of airmanship excellence is the best way to derive maximum enjoyment from personal flying. This pursuit must be driven by a desire to be the best aviator one can be. The pursuit of airmanship excellence must be continuous and it must be continued throughout an airman's entire flying career. If a personal flyer does this, he will be able to count himself among a truly elite group of aviators and he will achieve a level of enjoyment that will make the personal-flying value proposition very attractive to him.

I offer this chapter as an example of how the pursuit of airmanship excellence can provide the rewards every current and future pilot is looking for. Those rewards include expanded, unique personal-mobility options; an enhanced lifestyle; increased professional productivity; ever-expanding personal capabilities; a new perspective on life; the satisfaction that comes from mastering new challenges and recognition as a member of a very select community. This chapter will also provide you with enough of my aviation background to establish my bona fides as someone who may know what he's talking about when it comes to the pursuit of airmanship excellence. However, I in no way lay claim to being the final word in this area. As I have and will acknowledge throughout this book, the airmanship excellence that I pursue, and the one that I urge you to pursue, is the result of over a hundred years of contributions and sacrifices that were made by dedicated aviators and other aviation professionals who were pursuing airmanship excellence.

The following narrative of my pursuit of airmanship excellence is drawn primarily from my memory. Now, we all know about the reliability and unreliability of human memory. So, I must beg your indulgence for any lapses, fuzziness or unintentional distortions of those memories of mine that I recall below. I offer them primarily for the purpose of illustrating how one airman has progressed through his life-long pursuit of airmanship excellence.

I started my personal pursuit of airmanship excellence over fifty-seven years ago. I've achieved it many times. I say "many times" because although you may have achieved airmanship excellence in the past, it doesn't mean you are achieving it now. One must continuously exert the discipline and effort to acquire and maintain the airmanship knowledge and skills required by the type of flying one is doing. Senior airline captains can achieve and maintain airmanship excellence, and so can personal flyers.

1950s

I was born a couple of years after the end of World War II (1947). By the time I was able to begin to understand what was going on around me, TV was in its early stages of insinuating itself into every American household. Also, by that time the Cold War was really spooling up with a propaganda campaign that took perceptive advantage of this new communications tool (as well as the burgeoning movie industry). Therefore, I was exposed at a very early age to the plentiful WWII and Cold War propaganda movies that were frequently run as programming on the three (count them) stations that we could receive out in the farm country of northwest Indiana, or I saw them at the movie theater.

Many of these movies featured pilot heroes and the airmanship cultures of WWI, WWII and the Cold War. Therefore, I was unwittingly being indoctrinated into an airmanship culture that started with the Wright Brothers and matured during WWII (a span of less than 50 years). I'll have to admit that, like most of the boys in my cohort, I was strongly drawn to the airmanship cultures portrayed in these movies.

Although any reasonably intelligent, stable and coordinated person can learn to drive an airplane, not all of us are cut out to be pilots—far from it. I believe that for someone to take the risks and do all the hard work that is required to become a pilot, they have to be inspired by an intense innate desire to fly. I happen to be one of those people. If you're reading this book, I strongly suspect that you're one of them too. I believe that this fortuitous combination of continuous airmanship indoctrination through movies and television at a very early age, and my fundamental desire to fly that was somehow part of my personality, is what led to my life-long pursuit of airmanship excellence.

By the time I turned ten in 1957, I was watching everything I could find about flying on television and at the movies. I talked about being a pilot to the point that I started to annoy my mother who thought (with good reason at the time) that it was too dangerous for a young person to be involved in. I read everything about flying that I could lay my hands on—mostly magazines and newspapers because books on the subject of flying were not readily available in the rural community in which I lived. For you younger readers, that was back in the dark ages before the Internet and Amazon.

During this time, I built kites and model airplanes and flew them. I even ventured down the road a few miles to watch airplanes takeoff and land at the grass strip that served as Gary, Indiana's municipal airport. My relative proximity to the airport provided me with countless opportunities to watch light aircraft flying over my head in the traffic pattern for this airport. I have memories that I can still recall of lying on the grass in the backyard on a warm summer day watching a yellow Piper Cub complete circuit after circuit in the traffic pattern of this airport—the azure sky and puffy clouds forming a perfect canvas for what I even then recognized as a work of art.

I also frequently saw prop airliners like DC-3s, DC-4s and **Lockheed Constellations** as they descended from their enroute altitudes to make their approaches to Midway Airport from the southeast. This fairly constant exposure to aircraft flying over my head reinforced, I'm sure, my desire to learn to fly. By my eleventh summer, I had my mind firmly set on a career as a U.S. Air Force fighter pilot. I spent a good deal of my time learning about aviation and dreaming about my triumphs over my rival commie fighter pilots.

Thankfully my maternal grandmother always encouraged my flights of fancy. One of her cousins had been a WWII **B-25** pilot and she told me family stories about him and his flying exploits during and after the war. In fact, her son, my Uncle Bob, was also taken by those stories and used his GI Bill money from his service in WWII to learn to fly in Muncie, Indiana in the mid-1950s. These family stories, and the pictures my grandmother showed me of her cousin in his WWII pilot uniforms, served to reinforce my desire to be a fighter pilot when I grew up.

Early in the summer of 1958, my father announced that we were pulling up stakes and leaving the old family homestead near Merrillville, Indiana for the western frontier of the Chicago-metropolitan area. My two brothers and I were opposed to the move. By mid-summer, our resistance was mounting. Since I was the oldest and therefore the ringleader of the resistance movement, it was decided by higher-ups that I should be diverted while the move was initiated. This was accomplished with an invitation from my Uncle Bob to come on down to Muncie, Indiana for a couple of weeks so that he could take me flying in the airplane he had just bought.

I was easy. That was all it took for me to pretty much forget about the move to Illinois and focus instead on my first airplane flight. As it turned out, the reality was much better than the expectation. My uncle and my grandmother arranged for a formal flying lesson for me with my uncle's instructor. So, my first flight was a proper flying lesson with my grandmother coming along in the backseat of the **Piper Tri-Pacer** we used for the lesson. It was her first flight ever too! I remember being completely unconcerned for my personal safety, but I was worried about doing something stupid that would cause harm to her.

As I look back on it now, I can see how that very-early concern for being responsible for a beloved passenger set the benchmark for my attitude towards the thousands of passengers that I've been responsible for over the years. In fact, I always had a warm personal chuckle every time one of the old grizzled captains I flew with in the early days of my airline career counseled me about my aircraft handling techniques by telling me to, "Fly it like your grandmother is back there in coach, son. She'll appreciate it and so will I. After all, most of those passengers back there think I'm always flying the airplane and I don't want you to make me look bad."

My first flight was an amazing event, especially for an eleven-year old. Not only was I leaving the planet for the first time after intensely dreaming about it for a long time (to an eleven-year old), but I was taking my first flying lesson too. And the fact that I could show off what I knew about flying to my grandmother was an added bonus.

My first flight was pretty much what I expected after all my study of aviation. However, actually being up in the air and "in command" of a real live airplane was a thrill that I was indeed very fortunate to have given to me. I was definitely and irretrievably hooked. But it got even better. Over the next couple of weeks, this amazing first flight was followed up with two informal flying lessons with my uncle in his 1942 **Interstate Cadet**. He also spent several hours with me in the mellow Indiana-summer evenings going over the principles of flight, aerial navigation and aircraft mechanics. All of this was taking place less than 100 miles down the road from where Wilbur and Orville created the first airplane and wrote down the first principles of airmanship. I doubt that I thought about this fact at the time (although my study of aviation had by then included reading about the Wright Brothers), but as I look back, I can see that the seeds of my life-long pursuit of airmanship excellence were truly being planted in fertile soil.

I got to "pay" for my flying lessons by washing my uncle's airplane and helping him make a minor repair on it. At one point, I got to sit in the front seat of the Cadet by myself for over an hour while my uncle went off to retrieve a part. I think I flew around the world at least once in my imagination while I was waiting for my uncle to return. I discovered later that the ability to effectively "chair fly" is an essential element in the pursuit of airmanship excellence. I've used it throughout my flying career.

The three lessons I received over this two-week period were all based on an Airmanship 1.0 approach to flying. Civilian pilot training at that time did not give very much attention to the art and science of flight-risk identification, assessment and mitigation; nor did it put much emphasis on the human-factors aspect of airmanship. It concentrated instead almost exclusively on preparing new pilots for the FAA Private Pilot checkride. Unfortunately, Airmanship 1.0 training is still widely used in general-aviation today and it still suffers from these and other persistent shortcomings.

At the end of the vacation, I of course tried to convince my uncle to adopt me so that I could stay in Muncie and continue my flying lessons. Unfortunately, my pleas fell on deaf ears. But before we left for the new homestead in Illinois, both my uncle and his instructor told me that they were impressed with the aeronautical knowledge I had acquired and how quickly I picked up the flying techniques they had shown me. They enthusiastically encouraged me to pursue my dreams of becoming a pilot.

At the end of this magical summer adventure, I reported to our new home in northeastern Illinois. It only took the mention of the fact that there was a very active general-aviation airport only about four miles down the road, on top of my recent experience, to make me almost completely forget about my resistance to the move. And as icing on the cake, the recently commercialized O'Hare International Airport was a fairly easy jaunt via family car from the new home.

Although my mother still opposed my interest in aviation, my father came around and helped me to realize my dreams in any way that he could. For example, he would frequently take me on weekend-day excursions to O'Hare to watch the big birds takeoff and land and enjoy the heady aroma of burning jet fuel. He also, soon after our arrival in Illinois, took me to the local general-aviation airport (Roselle Airport, now Schaumburg Regional Airport) to find out about taking flying lessons. And he signed me up in a Boy Scout Troop because I had discovered that I could earn an Aviation Merit Badge in the program. I'm sure I didn't realize it at the time, but each of these opportunities turned out to be pivotal to my pursuit of airmanship excellence.

The trips to O'Hare, and a John Wayne movie that was playing on TV about that time—"The High And The Mighty"—got me interested in commercial aviation. Capt. Ernest K. Gann wrote the book the movie was based on, and the screenplay. Before that, I was focused solely on a military-aviation career. However, my plan was still to go after an USAF fighter-pilot slot first, and then maybe fly for the airlines after I retired from the military.

My time in the Boy Scout Troop not only trained me in leadership, survival and paramilitary skills, but a couple of the adult leaders in the troop (a WWII veteran and a Korean War Navy pilot) turned out to be important mentors to me. The former Navy pilot was my counselor for the Aviation Merit Badge and he gave me his Korean-War copy of the Naval Air Training And Operating Procedures Standardization (NATOPS) Manual.

Over the next couple of years, while I was in my "tweens", I read the manual countless times. It was my first formal introduction to a tightly organized, highly structured, professional aviation culture. I can't be sure that my exposure to this culture at an early age helped to set me on the path to pursuing airmanship excellence, but it probably did.

I know my early flight instructors always told me that I had a very professional approach to flying (they were mostly airline pilots and former military pilots). I believe that a big factor in the development of this professional attitude was my study of the NATOPS manual and my early assimilation of a professional-aviator culture.

By the time I was twelve, I had a paper route and I did odd jobs to earn money to feed my flying addiction. Of course, in the beginning I could accumulate only enough money for an occasional flight lesson. But when I turned fourteen, I could legally work in retail stores. Those part-time jobs generated enough income for me to take lessons more frequently.

I flew from Roselle Airport (now Schaumburg Regional Airport), Elgin Airport (now gone), Palwaukee Airport (now Chicago Executive Airport) and Elmhurst Airport (now gone) most of the time. I liked to fly the different types of aircraft offered by the flight schools on those airports, and I liked flying with different flight instructors, learning something new from each one of them. Private flight training at that time was unstructured and highly personalized by each flight instructor.

1960s

By the time I turned fifteen, I was well on my way towards my goal of becoming an Air Force fighter pilot. I was starting to accumulate the flight hours that I needed to qualify for an FAA Private Pilot Certificate, I was a leader in my Boy Scout troop, I was out for most every sport in school, my grades were excellent and I had concerned mentors helping me along the way.

One of the adult leaders in my Boy Scout Troop was a U.S. Air Force Recruiter. Given my strong interest in aviation, he offered to try to get me an appointment to the recently established USAF Academy. He felt that I fit one of the profiles the USAF was looking for in pilot candidates, and that if I worked hard I would have a good chance of getting an appointment given my early start on securing it. Back then, it wasn't unreasonable to think this was a route that might be available to me. However, just when things were off to such a good start, I suffered a major setback to my plans.

The setback came when I discovered that I didn't have 20/20 uncorrected vision. The recruiter and other adults helped me to ascertain that there was no way that I could get into USAF pilot training without it. I was encouraged to continue my pursuit of the Academy appointment with a shift in interest from flying to rocketry. Although I had a strong interest in space travel, it paled in comparison to my desire to fly airplanes.

After I spent a couple of weeks steeped in self-pity at having my dream vaporized in the blink of an eye, a couple of my mentors strongly encouraged me to look into becoming an airline pilot. This was the early 1960s and the airlines were just coming into their own with the first widespread use of jet transports. I could regularly see them arriving at, or departing from, O'Hare. My mentors also helped me focus on the fact that my real desire was to fly and there are many ways to scratch that itch.

My mentors introduced me to an American Airlines pilot who lived in our neighborhood (I'm sorry, but I can't recall his name.). He gave me a considerable amount of his time briefing me on the life of an airline pilot, the rewards for being one and how one could capture one of those highly coveted jobs. The critical piece of information he gave me though was that for some airlines (like United Airlines) you didn't have to have 20/20 uncorrected vision. It had to be, as I recall, no worse than 20/40 uncorrected and 20/20 corrected. My vision was 20/30 uncorrected and 20/20 corrected. It wasn't long before I reluctantly shifted my career goal from military pilot to airline pilot.

When I turned sixteen, I soloed for the first time (the minimum age) in a 1946 **Aeronca 7AC Champ** from a farmer's grass strip near Elgin, Illinois. The flight instructor who soloed me was Chad Koppie. At the time, Chad was a Delta Airlines first officer. In addition to polishing up my basic stick and rudder skills, Chad also acted as a mentor and provided me with excellent guidance in my pursuit of an airline-pilot career. In fact, I continue to consider him to be one of my mentors and I still have the good fortune to be able to sit down and have a cup of coffee with him occasionally.

After Chad soloed me, I continued to take instruction from him in his Champ for a few months until he decided to sell it. Since Chad's Champ was no longer available to me, I thought the best plan was for me to buy a Champ of my own to continue my flight training in. I found one at the Sandwich, Illinois airport. I convinced a banker to loan me the money to buy it (with my Dad cosigning), and Chad signed me off for solo cross-country flight. A friend dropped me off after school at Sandwich Airport to pick up the Champ on a very cold, wintry day.

After a considerable amount of effort, I finally got the Champ started. I had to hand-prop it because it didn't have an electrical system (and thus no starter). After getting the engine started and warmed up, I flew the Champ to Roselle Airport where I intended to keep it. What a thrill to be flying my own airplane by myself. At the tender age of sixteen, I had acquired another addiction—aircraft ownership. So far, I have recovered from, and subsequently backslid into, aircraft ownership at least eight times.

The fact that I was an aircraft owner who rented a tie-down space and bought fuel and maintenance services from the FBO (fixed-base operator— Roselle Beechcraft) at Roselle Airport opened several doors for me. I was able to get a part-time job at the airport washing airplanes, pumping avgas, mowing the grass and doing janitorial chores. This income stream, meager though it was, when added to my wages from working part-time at the local Jewell store, was enough to allow me to fly my own airplane three or four times a week. And I enjoyed the pleasure of being around airplanes and pilots for several hours a week while I was performing my part-time chores around the airport.

I also got an employee-discount on flight training that helped me to accumulate the flight time and log the training I needed for my immediate airmanship goal—my FAA Private Pilot Certificate. But more importantly, from an airmanship-development point of view, the real value in learning to fly at the Roselle Beechcraft flight school was the quality of the instructors that I was able to fly with. They ranged from a WWII RAF Spitfire pilot, to three young aviators who were on the verge of being hired by the airlines (Bob Lussow, Jim Raymond and Norm Pesch—they all were hired by the airlines, Bob by Delta, Jim by North Central and Norm by Eastern) to two moonlighting airline pilots (John Rolfing of TWA and Mike Cosby of United).

The culture at the airport was definitely an Airmanship 1.0 culture, but it was biased towards the way professional pilots were practicing it at the time. This served to indoctrinate me into the best Airmanship 1.0 practices of that era. Those best practices included: concern for my passengers (their safety, comfort and enjoyment), rigid flight discipline, thorough preparation for anything aeronautical I was about to undertake, personal integrity, proper aircraft handling and the use of the command authority that is bestowed upon every Pilot In Command. This brand of Airmanship 1.0 culture was not typically found in the general-aviation flight organizations of the time.

I also had the great good fortune to become friends with another young aviator who was pursuing a professional-pilot career. He was Bob Lussow's younger brother Rich. Rich and I provided peer pressure for each other and flew together a lot. Rich went on to fly for several airlines and capped his career by retiring from United Airlines. He now works for the FAA as an aviation inspector who oversees the simulator-training program of a major airline.

Shortly after basing my Champ at Roselle Airport, a flying club also based on the airport offered to buy it from me. The deal included a membership in the club. Since I would soon need more capable aircraft to continue my flight training, I sold the Champ to the club and used it and the club's **Cessna 120** and **Cessna 182** for the remainder of my private-pilot training.

The flying club (Elgin Flying Club) and the restaurant located on the airport provided very fertile ground for the growth of a strong Airmanship 1.0 culture. I thoroughly enjoyed the social aspects of that culture. I learned very early that if you want to have a vibrant airmanship culture of any kind, the members of that culture have to interact formally and informally on a frequent basis. And they need a place to meet to carry on that interaction. The building with the old restaurant in it has been torn down and replaced by a modern terminal building with "Pilot Pete's Restaurant and Bar" in it. Local pilots still use it as a gathering place for social and airmanship-related activities.

I progressed fairly rapidly through my pilot certificates given my time and financial constraints. I passed my FAA Private Pilot checkride just after I turned seventeen (the minimum age) and I earned my FAA Commercial Pilot Certificate, Instrument Rating, Multi-engine rating and Certified Flight Instructor Certificate shortly after turning eighteen (the minimum age). John Rolfing (TWA) and Mike Cosby (UAL) provided virtually all of my advanced training. John and Mike, since they knew they were grooming me for a career as an airline pilot, worked hard to make sure I had the knowledge, skills and attitudes that I would need to be successful in that professional-airmanship culture.

Before the ink was dry on my Certified Flight Instructor Certificate, I was promoted from line-boy to flight instructor by Dick Leach who was the manager of Roselle Beechcraft flight operations at the time. Dick was a former airline pilot who had to leave his airline job due to a medical problem. Fortunately, he could still qualify for an FAA medical certificate that allowed him to fly general-aviation aircraft professionally. I was very fortunate in that this promotion allowed me to step into a full-time flying job immediately after graduating from high school in 1965.

I was soon flying virtually every day giving flight instruction and ferrying aircraft around the country for the company. I checked out in many aircraft types and learned a lot about flying and teaching people how to fly. My mentors—Chad Koppie, John Rolfing, Mike Cosby and Dick Leach—helped me to get the most out of that experience and were always ready to rein in my youthful exuberance whenever it threatened to get me into trouble (both in the air and on the ground).

I was definitely a typical product of my environment. As a teenager and twenty-something, I suffered from the sometimes-fatal malady contracted by young men throughout the ages—an unrealistic evaluation of my vulnerability. I was, I thought, immune to the tragic mistakes made by other aviators. I had a very virulent case of "it can't happen to me".

I certainly knew about those mistakes. They were well covered in all of the contemporary general-aviation publications and the FAA was even then pushing hard for aviators to take every means necessary to avoid them. My macho attitude and youthful false sense of invulnerability led me to take risks that I definitely shouldn't have. I believe this was due in large part to the fact that the airmanship knowledge and techniques that all of us Airmanship 1.0 aviators needed to methodically identify, evaluate and mitigate flight risks were not then available.

So, I thank my lucky stars and my mentors for somehow helping me to survive this very-high-risk phase of my airmanship development. And I definitely don't recommend that approach to anyone. I'm lucky to be here today and there's no valid reason for any aviator to go through a high-risk phase in the pursuit of airmanship excellence. There are better ways to do it.

My approach to airmanship excellence was also informed by other airmanship-related experiences that I had during this period. It wasn't very long after I started flight instructing for Roselle Beechcraft that Bob Lussow, Jim Raymond and Norm Pesch were hired by the airlines as pilots. This left a pretty big gap in our instructor core and Pesch's departure left us without a chief instructor. Dick Leach offered me that job and I took it. My responsibilities included recruiting new students for the flight school and managing the flight-training program that we offered to the public.

As you will see, an important aspect of airmanship excellence is increasing the number of pilots who practice Airmanship 2.0. This includes converting existing pilots, but more importantly, recruiting new aviators to the Airmanship 2.0 approach to aviating. In my role as chief instructor, I conducted dozens of meetings around the community wherein I recruited many new student pilots. The lessons I learned then about recruiting pilots have served me well over the subsequent decades in many diverse endeavors.

I was also very fortunate to have been chief instructor when Roselle Beechcraft purchased one of the very first **Frasca** simulators that had been specifically designed for general-aviation flight schools. Although the military and airlines were using rudimentary simulators at the time, very few civilian flight schools had them. Unfortunately, this is still the case today.

As the chief instructor, I was tasked with developing an FAA Instrument Rating course based on the Frasca simulator since no such program was available on the market at that time. This introductory flight-training-program-development experience set me on a lifelong course of designing and implementing flight-training programs. The knowledge and skills I have acquired over the years in this area have impacted the way I approach airmanship excellence in many ways.

At this time, at the age of eighteen and with only around five hundred hours of total flight time, I had a full load of students. I flew every day that the weather would allow with these students. Most of them were just starting out as pilots. So, my job included indoctrinating them into our airmanship culture. It was also part of my job to explain learning to fly to prospective students who came to us for information. I worked with scores of new and prospective pilots who of course had their own ideas of how to approach flying. This experience laid the foundation for my lifelong study of pilot cultures and how one goes about creating and changing them.

Even though I had, for pragmatic reasons, changed my career goal from military pilot to airline pilot, my heart was still in flying for the military. As I mentioned earlier, my cohort had been heavily influenced by cold war propaganda and I was, I felt, a patriotic warrior at heart. My experience in the Boy Scouts reinforced this orientation throughout all of its programing. In junior-high school, I had even practiced short-order drill on a weekly basis under the command of a Marine drill sergeant who served as my physical-education teacher. And of course, the Boy Scouts were long on weapons training and field tactics and the majority of our adult leaders were WW II or Korean War veterans. In addition, during this time, the Vietnam War was beginning to spool up. I, and most of my friends, anticipated military service soon after graduating from high school.

Like most teenage boys since the dawn of time who have been indoctrinated into a warrior culture, I wanted to go to war. I just didn't want to go to war in a non-flying capacity if I could help it. Shortly after high-school graduation, I contacted an Army recruiter because I had heard that the Army was recruiting helicopter-pilot trainees without college degrees as was required by the Air Force and Navy. I had just started my college training. I hoped that since I had all of my professional-pilot qualifications, I might be able to get a waiver on my eyesight. I knew that once a pilot got out of initial military pilot training, his eyesight only had to be correctible to 20/20.

The Army recruiter set me up to go through the Army's pilot-screening program that consisted of a thorough physical exam and extensive aptitude testing. After completing the program, the recruiter told me that I had passed everything with flying colors except for the eye exam. I had expected this and laid out my arguments for a waiver of the vision requirements based on my professional-pilot qualifications. The recruiter sent my request up his chain of command.

A couple of months later, the word came back down that the Army wasn't interested in granting a waiver at that time. However, I was offered a "guaranteed" position as a Huey crewchief/door-gunner. I wanted to accept the offer because I believed that if I could join the Army and get involved in Army aviation, I could get around the requirement for uncorrected 20/20 vision. My mentors convinced me to decline the offer. Thank God I listened to my mentors.

During my first year in the aviation business, I accumulated over one thousand flight hours and learned many lessons about how to properly practice the art of airmanship. By the end of that first year, when I was nineteen, I secured an interview with United Airlines for a flight officer (pilot) position. After the first series of interviews, I was run through United's pre-employment gauntlet. This process was focused on airmanship qualities, knowledge, experience, aptitude and medical condition. At the completion of this process, United offered me a job as a new-hire second officer. I, of course, eagerly accepted the offer. However, there were 2 problems—I couldn't qualify for an FAA Flight Engineer Certificate (this was required for the entry-level position) until I reached the age of twenty one (that was about a year-and-a-half away), and the job offer would be rescinded if I was called up for, or volunteered for, military service.

Waiting until I turned twenty-one was the only way I could solve the first problem. The second problem was a little more difficult to solve. By this time, the Vietnam War draft was heating up and I could feel the flames lapping at my backside. If I enlisted in the military, even if I was able to get into pilot training, my golden opportunity to fly for the airlines would be forfeited. If I did nothing at all for the next two years, I risked having my number come up in the draft. If I got drafted, not only would I miss securing the United slot, but I would also most likely serve in some non-flying capacity in the military. The only other option I could think of was to join the National Guard, but young men who had good connections were taking the available openings in the Guard. As it happily turned out, I had one of those good connections.

The connection was Thad Slaughter. Thad was a few years older than me and he instructed part-time at Roselle Airport. Thad was also a well-respected member of the Illinois Army National Guard who served part time as a maintenance technician with the Army helicopter outfit based at Midway Airport. Luckily for me, Thad was a good friend of the outfit's First Sergeant.

Thad invited me to visit the unit at Midway during a drill weekend. At the time, the Guard was more like a flying club than a military unit. Several of the flying officers were airline pilots in their real lives and most of the enlisted men were professionals of some sort. I felt at home there and asked the First Sergeant if there were any openings in the unit. He asked me when I could leave for basic training. Thad had obviously greased the rails for me.

I immediately checked with United Airlines to see if they were OK with me joining the National Guard. Without anticipating it, I was the recipient of another big payload of good luck. As it turned out, there was a law on the books that prohibited companies from denying employment to, or terminating the employment of, members of the National Guard. United told me that they would honor their job offer and wished me well. The only thing they asked is that I try not to be in Army training when my United class date came up in 1968.

For the next six months or so before I left for basic training in the Army, I was the chief pilot of Chicago Steel Erectors' corporate flight department. I ran the department and flew their new Beechcraft Baron (a light-twin-engine aircraft). Bill Medley, the CEO, was a private pilot to whom I occasionally gave instrument instruction in addition to flying him, his family and his employees around the country in the Baron. Bill, George Northam and I bought a 1942 **Boeing-Stearman Model 75** open-cockpit biplane. I really enjoyed flying the airplane and taking people for rides in it. I felt like a real barnstormer.

While I was flying for Bill, he became another mentor to me—primarily in business. In fact, he tried to discourage me from my professional flying dreams and instead offered to teach me his business. Although I was very interested in the construction business and really wanted to take Bill up on his offer, the pull of professional piloting was too strong. When I had to leave Bill's employ to complete my Army training, he replaced me with a series of part-time pilots.

I shipped out to Army Basic Training at Ft. Campbell, KY in mid-winter 1967. I soon discovered that Kentucky winters and the red clay of the base didn't provide for the most pleasant training environment. But since Ft. Campbell was the home of the 101st Airborne Division, I at least had **C-130s** almost constantly flying over my head as I acquired the basic-combat skills the Army thought I should have.

The big threat the drill sergeants held over our heads to motivate us was their professed belief that all of us would be in Vietnam in the very near future, even those of us who were in the Guard and Reserves. At first, I took their threats with a grain of salt, believing that I would return to my home unit at Midway as planned after I completed my advanced training. However, before the training was half over, most of us who were in the Guard and Reserves were beginning to think the drill sergeants might be right. Rumors were starting to circulate that President Lyndon Johnson was going to call up the Guard to augment the rapid troop buildup in Vietnam.

After basic training, I moved on to advanced training at Ft. Eustis, VA. In 1967, the fort was the home of the U.S. Army Transportation Corps. I was enrolled in the MOS 67N20 course. It would qualify me to serve as a Huey crewchief/door-gunner. That's right, the same slot I had turned down a couple of years earlier when the regular Army offered it to me.

The course also trained me in fixed-wing-aircraft maintenance and reciprocal-and-turbine-aircraft-engine maintenance. This training provided me with an excellent foundation for acquiring airmanship knowledge about the aircraft I would fly throughout my career. That's important because as you'll see, "Knowledge of Aircraft" is one of the five pillars of knowledge in the Kern Airmanship Model.

The training program was excellent and it even included an orientation hop in a **Bell H-13 Sioux** helicopter (think Whirly Birds and MASH). The warrant officer who gave me the ride soon turned the orientation flight into a flying lesson when he learned that I was a professional pilot in civilian life. We formed a fast friendship because he was about to separate from the Army and wanted very much to get a job as an airline pilot. I traded information and contacts with him for more flying time in the Bell.

About a month before I finished my maintenance-training program, my new friend the warrant officer told me that he had been talking to a friend of his in personnel and he thought that I should apply for Army flight school. The personnel guy thought that I would probably be accepted based on the facts that I was now in the Army, about to finish my basic technical training and I had over one thousand hours of pilot time and all of my professional-pilot certificates. By the time this opinion was offered, I was becoming pretty "gung ho" about military life in a wartime army. The warrior mentality that I had been indoctrinated with in my earlier years, and the intense exposure to military culture that I had experienced over the past several months, had turned my head away from airline flying back to military flying.

After considerable soul searching, I decided that I really wanted to go to war and I wanted to do it as a helicopter pilot. I think that a big factor that led to this kind of thinking was all of the real-war stories I was hearing from recently returned Vietnam vets who served in helicopter units while they were in-country. I followed up on my friend's suggestion. It wasn't long before I was going through the screening process for Army-pilot training for the second time. In retrospect, the risk-reward assessments that a nineteen-year old can make are truly amazing. The results of the screening process were the same as the one I had gone through earlier—I was more than qualified for Army-aviator training with the exception of my uncorrected vision. Once again, an appeal was sent up the chain-of-command for a waiver on my vision, and once it again it came back down with a "not-at-this-time" answer. I resolved to forget about a military-aviation career and focus instead on going to work for United Airlines as a pilot. My plan was to serve out my six-year commitment with the Guard as a part-time Huey crewchief/door-gunner while I was flying for United Airlines full-time. I would keep my fingers crossed when it came to rumors of deployment to Vietnam.

While I was in Army training, I dreamed about starting my own aircraft-charter business. My experience with the charter department at Roselle Beechcraft led me to believe that one could launch a successful charter operation and gain sufficiently interesting market share if one were to go out and sell light-aircraft charter. At that time, light-aircraft-charter marketing consisted of Yellow Page ads and word of mouth. During my six-month Army training stint I worked through plans on how to pull off my first start-up venture. I used the business-planning process as a diversion while I was doing uninteresting things like marching and KP.

When I returned from my Army training, I approached Bill Medley (my former corporate-pilot boss) with my ideas on starting an aircraft-charter business. I also talked with a former student of mine, Joe Palazzo, who was a successful entrepreneur and private pilot who owned a new **Beechcraft Debonair** (a top-of-the-line single-engine airplane). It wasn't long before the three of us had formed a partnership. Bill and Joe leased their airplanes to the new venture, they provided the needed start-up capital, they mentored me on the ways of entrepreneurship and I did everything else.

We soon had All-States Aviation up and running. To everyone's surprise, we started making money almost from the day we opened our doors. By the time I turned twenty-one, the business had grown to five airplanes, fifteen pilots and three administrative staffers. I was amazed at how eager people are to give you their business if you ask them for it and prove that you can deliver on your promises.

This experience provided me with the opportunity to learn a lot more about airmanship because I had to personally write the FAA-approved flight manual for the operation. Mike Cosby loaned me his United Airlines Flight Operations Manual to use as a framework for the one I was writing, and I got a lot of help from another mentor of mine, Theo Moore, who was an FAA air-safety inspector.

I had met Theo a few years earlier when I transferred from the Boy Scout troop I had been in since grade school to a Boy Scout Aviation Explorer Squadron that Theo was serving in as an adult leader. Theo, an Army aviator before joining the FAA, had also strongly influenced my approach to airmanship up to that time and he continued to do so for several years thereafter.

In writing the flight-operations manual, creating the business's operating policies, recruiting and training pilots and running the business on a day-to-day basis, I received an intensive course in how to structure and run what I now call an Airmanship Development Support Organization (ADSO). I also discovered in myself a keen interest in the art of airmanship and how one goes about pursuing airmanship excellence.

While I was running the charter operation, I decided to explore the possibility of offering helicopter-charter flights in the Chicago area. I had always been interested in flying helicopters and that interest was intensified with my involvement with them in the National Guard. I figured that if I was going to launch a helicopter service, I better learn how to fly one.

I was getting a little "bootleg" time in my Guard unit's helicopters, but I decided to pursue an FAA commercial-helicopter-pilot certificate through a civilian flight school. At that time, there weren't many flight schools in the area that offered helicopter training, but I found one that was located on Joliet Regional Airport in Joliet, Illinois. It wasn't long before I soloed in a **Hughes 300**, the same helicopter the Army was using to train its new warrant officers at its Army Primary Helicopter Training Center in Mineral Wells, Texas. However, shortly after I soloed in the "300", I suspended my helicopter training. United Airlines had informed that it was time to report to the United Airlines Pilot Training Center in Denver.

I celebrated my twenty-first birthday in May of 1968 and left Chicago in July for United's training center in Denver. A two-day drive in my brand-new, dark-blue Volkswagen Beatle gave me ample time to reflect on the whirlwind three years I had just experienced since graduating from high school and the new experience I was just hours from jumping into. The past three years had qualified me to take a shot at playing in the big leagues. Now I was going to find out if I really had what it takes to make my way at the top of professional piloting. I couldn't wait to find out.

Chapter 1: My Pursuit of More Enjoyable Personal Flying

There were five former military pilots and two civilian pilots in my United Airlines **Boeing 727** new-hire class. I was the youngest at twenty-one and the "old man" was thirty-seven. It started out with eight hours of classroom training five days a week for the first couple of months. In our "off" time, we spent countless hours poring over our manuals and FAA regulations and sitting in front of our "paper tigers" (full-color, life-sized pictures of the Boeing 727's control panels). We were also subtly encouraged to participate in United's version of an Airmanship 1.0 culture by challenging each other on every detail and then demanding that the smart guy with the answer cite the correct reference for that answer.

At the time of my arrival at United, the flight-operations culture was decidedly military in nature. Most of the managers, instructors and pilots came from a military-aviation background. Military-like appearance and bearing were required. Discipline was tight. Procedures were detailed and complete. There was almost reverential deference towards the captains by the first and second officers. And the pilots who had risen from the ranks to become flight managers were considered to be of the highest authority. Captains' decisions were rarely challenged. New pilots were strongly encouraged to learn from these old salts and the captains were required (it was in the flight manual) to mentor their subordinates (first and second officers).

But above all, the culture demanded that every one of its members practice continuous airmanship development. In other words, the pursuit of airmanship excellence was a core value of the culture. Other core values included professional integrity, fairness, concern for passengers and crew, safety, decisiveness, courage, resourcefulness and a superior knowledge of the art of airmanship.

The behaviors deemed appropriate to realize these core values were clearly laid out in the flight-operations manual and continuously modeled by the culture's leaders. Frequent recognition was awarded to those who exemplified these behaviors and appropriate penalties meted out to those who didn't. Friendly competition was encouraged among crewmembers to see who the best pilot in the cockpit really was. All of this led to what I consider to be the pinnacle of Airmanship 1.0 culture. This type of culture existed at most of the major airlines at the time (the late 1960s), but not, in general, in other sectors of the commercial and private-aviation worlds.

After the approximately two months of ground school wherein I was afforded a very thorough education on the intricacies of the Boeing 727, we moved on to about a month of simulator training. This was my first-ever formal exposure to flying with two other professional pilots in a highly structured flightdeck environment. Prior to this time, my cockpit-management skills had to be applied only to me as a single-pilot, or to a student if I was giving instruction. Now I had to work very closely with two more-experienced aviators who expected from me nothing less than perfect performance. The good news was that the captain and first officer in my training crew were already experienced United pilots who were moving up in aircraft and/or moving up a seat. They both took their responsibilities as professional mentors seriously and were always patient and helpful until I got the hang of things.

It didn't take long for me to realize that flying with a crew opened up a whole new area of airmanship that I hadn't really been exposed to before with the exception of the WWII bomber-command movies that I still enjoyed watching. I was fascinated with the prospect of learning everything I could about this new professional world that I planned to spend the next thirty-nine years of my life in. It was an environment that quickly became comfortable to me.

As I look back on it now, I don't know how I found the time given my very-heavy study-and-training schedule at the United Training Center, but I managed to carve out enough time in my weekly schedule to drive to the Black Forest Glider Port north of Colorado Springs several times to learn to fly gliders. Over the course of my stay in Denver, I obtained my FAA Commercial Glider Pilot certificate. The mastery of powerless flight was not only exhilarating, but it also added to my bag of airmanship capabilities.

At the end of the United simulator-training course, I passed the written exam, oral exam and checkride for my FAA Flight Engineer-Turbojet Certificate. But the best part of the United new-hire training program was the actual flight time I got in a Boeing 727. The deal was, if you passed the Flight Engineer course you got two hours in the right seat of the "Three-Holer" simulator practicing takeoffs and landings and then you got to do three takeoffs and landings in the real airplane down at Colorado Springs.

Up to then, the biggest airplane I had flown was a **Beechcraft D-18** Twin Beech. The D-18 was a conventional-gear (tailwheel) airplane powered by two reciprocating-radial engines of 250 hp each. It accommodated six passengers with a maximum takeoff weight of 7,500-pounds and its maximum speed was 225 mph. The Boeing 727-100 that I was going to fly was a modern jet transport powered by three Pratt & Whitney JT8D-1 jet engines with 14,000 pounds of thrust each. It accommodated around 150 passengers with a maximum takeoff weight of 170,000-pounds and its maximum speed was 550 mph. Needless to say, I eagerly anticipated the opportunity to see if I was up to the challenge.

A United flight instructor rounded up all seven of us in my class and we went out to the flightline. On the way from the Training Center to the flightline, we drew straws to see you would fly first. I got the long straw and was designated copilot for the first leg of the flight. My six fellow trainees would trade off in the engineer seat and take their turns in the right seat after I completed my three "bounces". The good news about being first to fly was that I got to make the takeoff in Denver and then fly the airplane down to Colorado Springs where I would make two more takeoffs and three landings. This provided me with more Boeing 727 flying time than all of my classmates except the guy who won the chance to fly back from the Springs to Denver. The bad news was that I was going to be the first of us to have his 727 flying critiqued by the rest of the class as well as the instructor-pilot.

As it turned out, with coaching from the instructor, I was able to make three decent landings. My classmates who came from a military background congratulated me on my performance and expressed their surprise that a "little airplane driver" could fly the 727 so well his first time out. I basked in their praise, but I knew I had a long way to go before I could call myself a "real airline pilot". In fact, I had one more obstacle to overcome before I could count myself among the elite of professional flying. That hurdle was an initial line check that I would have to pass after reporting to my first United Airlines duty station.

Near the completion of United's new-hire school, a person from personnel met with me to go over my near-term-future options for crew bases. During the meeting, he told me that personnel was projecting that I would probably be eligible for a captain's bid before I turned twenty-three. The airlines were expanding rapidly at that time and the seniority list was highly upwardly mobile. That was a problem for me since the minimum age required for an Airline Transport Pilot Certificate (which is required to fly as a captain for an airline) is twenty-three. That meant that I'd have to wait until then to actually become a captain. As you'll see, things didn't quite turn out that way, but the personnel rep's comments certainly planted expansive dreams in my head that were later to be treated to a harsh dose of reality.

It's probably a good thing that I had to wait to make captain. I felt then that I'd be ready to command a jet airliner by the time I turned twenty-three. After all, what red-blooded young pilot doesn't think that he can fly anything with wings on it? And besides, I knew I would look really good with four stripes on my uniform sleeves and a bulging wallet. But as I reflect on it now, I'm convinced it would have been premature—literally—for me to take a captain bid at that age. I didn't get over my youthful attitudes about risk taking until much later. Today, I have to admit that if I got on a regional airliner and saw a very-young crew in the cockpit, I'd definitely turn around and get off. In my opinion, the minimum age for an airline captain should be somewhere north of thirty.

I left Denver in October of 1968 for New York City where I was posted as a Boeing 727 second officer flying out of JFK and LGA. I kept the "pedal to the metal" in the Beatle all the way from Denver to the Big Apple with only a short stopover in Chicago to visit family and friends. I couldn't wait to put on my shiny-new blue uniform and actually take my place as an airline pilot.

However, I never got a chance to put it on and fly a trip for the New York base. Immediately upon reporting to my Flight Manager at JFK, I got "junior-manned" into a temporary-duty assignment. I was told to turn around and go back to Chicago where I would serve as a 727 second officer based at ORD for at least ninety days.

This was more than OK with me because I had wanted ORD as my first domicile. However, because I was the junior man in my new-hire class due to my age, I didn't even have a choice of bases. When they got down to me, the only base left was New York. So, I did as I was told and drove back to Chicago to start my new career. As it turned out, I was able to stay on temporary duty at ORD for six months. This allowed me to draw temporary-duty pay the whole time. That was a very important boon to a starving first-year second officer on probation pay. At the end of the six months, I captured a bid to Chicago as a 727 second officer.

My first flights out of Chicago as a member of an airline crew were truly amazing. Prior to my first short flight on a 727 during my Flight Engineer checkride, I had been on an airliner (a 727) only once before (as a passenger). Now I was routinely jetting back-and-forth across the U.S., and I had the third-best seat in the house. I was also enjoying learning real airmanship literally at the elbows of the best jet-transport-aircraft pilots on the planet.

When I arrived on the line for United in late 1968, the vast majority of the captains were current (serving in the Guard or Reserves) or former military pilots. About two-thirds, as I recall, of the first officers (copilots) were from the same background. My generation of airline pilots was comprised of approximately two-thirds military and one-third civilian pilots.

Since I only flew with captains and first officers, I was exposed to mentors who were primarily from a military-flying background. The flightdeck atmosphere was decidedly military in nature. I was able to fit in because by this time I had been serving in the National Guard for over two years. My role in the Guard as a crewchief was not all that different from my job as a flight engineer (second officer).

Chapter 1: My Pursuit of More Enjoyable Personal Flying

The type of airmanship being practiced at United at that time, what I call Airmanship 1.0, was an amalgam of military and airline airmanship that had by that time progressed about as far as it was going to before Airmanship 2.0 evolved out of it at the airlines and military in the early 1980s. The captain was still the king of the cockpit and first and second officers rarely challenged the captain's decisions. In fact, many captains at that time did not actively solicited input from other members of the crew. This is a defining characteristic of Airmanship 1.0.

Nevertheless, I was having a great time even though the airmanship learning curve I was on was definitely steep. I thought I knew a lot about flying airplanes before I started flying for United. But these pros, who were at the top of their game, taught me how to operate in a multi-crew environment within the very-demanding world of airline flying.

Not long after I completed my probationary year with United, which resulted in an approximate doubly of my flight pay, I purchased two airplanes. The first was a **Cessna 177 Cardinal**. It is a four-seat, single-engine personal airplane. The second was a **Grumman American AA-1 Yankee Clipper**. It was a recently introduced, leading-edge, two-seat, single-engine sport airplane. I enjoyed flying both airplanes around the U.S. for my business-and-personal-mobility needs.

1970s

In 1970, I decided to take a 727-second-officer bid in Denver. My primary reason for transferring from Chicago was to live in the mountains. I had grown up in the flatlands of Indiana and Illinois. So, my first exposure to the Rocky Mountains during United Airlines new-hire training at United's Denver Training Center was truly inspiring. Since that training in 1968, I had planned to take up residence in Denver as soon as I could. The 727-second officer bid was my first opportunity.

The United airmanship culture I experienced in the Denver base was markedly different from that in Chicago. Of course, the Denver crews followed standard operating procedures and adhered to United's policies just like the Chicago crews, but the overall atmosphere among the pilot group was much more laid back. I flew my personal airplanes to Denver and flew them extensively throughout the area. I traveled back to Chicago for my National Guard drill weekends and occasionally drilled with the Army Guard unit at Buckley Air Force Base near Denver.

However, after a little more than a year at United's Denver base, I decided to take a **DC-8**-second-officer bid in Chicago. The pay increase associated with the bid was significant and I had always wanted to fly a big four-motor jet

29

transport. I completed my DC-8 training at United's Denver Training Center and headed back to Chicago.

Around 1972, the predictions made by my Army drill sergeants during basic training in 1967 looked like they were going to come to pass. The Guard Bureau informed my unit that we should start preparing for deployment to Vietnam. This bit of news obviously had significant implications for my unit and for me.

Since the Vietnam War had been going on for some time by this point, a large number of Army pilots had returned from tours of duty in Vietnam. These aviators, for the most part, had no desire to return. So, they were not joining the Guard. Also, the WW II veterans in our unit were beginning to retire from the unit, some due to age and others because they didn't want to deploy to Vietnam at that point in their lives. Therefore, my unit was very short of pilots.

I was determined not to go to war as a crewchief/door-gunner. So, I renewed my efforts to get into Army flight school. I wasn't getting very far until a particular Saturday evening during a drill weekend. It was part of our culture for the unit's officers and enlisted men to gather in the officer's club at the end of the day for relaxation and conversation. The National Guard at that time was a very egalitarian organization, and it was not unusual for us to socialize across the ranks.

It wasn't long before the conversation turned to how we were going to get up to strength for our impending deployment. One of the WW II vets who flew **B-17s** over Germany during the war told us about how he ended up as a United States Army Air Forces flight officer. He related how he got into Army pilot training because he had civilian flight time prior to the start of the war. It wasn't long after the war started that the Army brass realized that the thousands of new pilots being turned out by the Army with the rank of second lieutenant would offer considerable competition for promotions after the war. They therefore shrewdly invented the flight-officer program to take care of that problem.

After WWII (in 1947) the U.S. Army Air Forces morphed into the United States Air Force and the flight-officer program was ended. However, the U.S. Army retained several aviation assets and therefore needed pilots. These pilots were called Army Aviators. In 1953, the U.S. Army Warrant Officer Flight Program was created. The Army brass created the program for the same reasons the U.S. Army Air Forces set up the flight-officer program in WWII. It limited the number of regular officers competing for the top spots in the Army hierarchy. This program was still in place in the early 1970s and in fact was providing most of the Army pilots who flew in Vietnam.

Chapter 1: My Pursuit of More Enjoyable Personal Flying

Our WW II vet suggested that the civilian pilots in our unit (there were many including about a half dozen airline pilots) who wanted to transition from the enlisted ranks to the pilot ranks before we were deployed might want to consider petitioning the Guard Bureau to institute a program that would allow civilian-pilot members of the Guard to enter Army flight training based on their civilian-flying experience. I thought this was a pretty good idea. So, I organized the civilian pilots in our unit who wanted to do that and we started working on the powers that be to let us make the transition.

As I recall, it took less than a year to convince the Guard Bureau of the wisdom of this course of action. As it turned out, those of us with over two thousand hours of civilian pilot time who could make it through the Army's pilot-selection gauntlet were invited to volunteer as Army Aviators. And the best part of the deal for me was that all I had to do was pass an Army first-class flight physical to qualify for the program. I didn't need 20/20 uncorrected vision. Just vision that was at least 20/40 uncorrected and 20/20 corrected.

It wasn't long before the Army waived its magic wand I was anointed as an Army Fixed Wing Aviator with the rank of Warrant Officer-1. The only rub was that since I hadn't attended an Army fixed-wing training course, I couldn't fly any of the fixed-wing or rotary-wing aircraft in my unit. Sending me through the US Army rotary-wing transition course at Ft. Rucker, Alabama rectified this problem.

During my six-month training program at Ft. Rucker, I transitioned into helicopters in the **Bell OH-58 Kiowa** and then moved on to tactical and instrument training in the **Bell UH-1 Iroquois** (we always referred to is the Huey). The training was highly professional and compared favorably with my airline training. I was taught the Army's version of Airmanship 1.0 that was almost identical to the Airmanship 1.0 I was practicing at United.

The Huey, as flown by the US Army, is a three-crew aircraft— pilot/aircraft-commander, copilot and crewchief/door-gunner. I was trained to be the aircraft commander, which is equivalent to the position of captain on the airlines in terms of responsibilities and command authority. I was struck by how similar the way Army aircraft commanders ran their crews was to the way United captains ran theirs. The captain's/aircraft-commander's decisions were rarely challenged by the other crewmembers and input from other crewmembers related to those decisions was seldom sought. These two behaviors are typical of an Airmanship 1.0 culture.

Upon completion of Army flight school, I returned to my Guard unit at Midway Airport to take over as the unit's training officer and its most-active instrument instructor. Several of the unit's pilots did not have helicopter-instrument ratings and the Army had just handed down an edict that all Army

31

pilots deploying to Vietnam would have one. Because I had a lot of time off from my day job at United Airlines, I gave a ton of helicopter-instrument training to my fellow Army Aviators.

My role as training officer afforded me with the opportunity, that I took full advantage of, to learn the Army's approach to training. And my job as an instructor-pilot provided me with a lot of hands-on experience in using the Army's system. It was a much-different type of training environment than the one I had grown up in during my earlier stint as a civilian flight instructor. The Army's approach to flight training was almost identical to the airline's with the exception of the use of simulation. The Army now uses simulation as much, or more, than the airlines do, but way back in the early 1970s, simulator time was not part of the Rotary Wing Transition Course.

Every training flight was preceded by at least a one-hour briefing and a debriefing of a similar length followed it. Trainees followed a highly structured, methodical syllabus. All trainees completed ground school before beginning flight training and training documentation was meticulous.

Back at United, the movement of my seniority number up the list was not progressing nearly as fast as the personnel representative had predicted when I was in new-hire school. When I turned twenty-three, instead of being able to get a captain's bid, I was still stuck in the back seat as a second officer. And the way my seniority was progressing, it looked like it would still be several years before I even got a first officer bid. The slow progression of my seniority was due primarily to the 1969 – 1970 recession and the resulting cutback in airline flying. But that didn't stop me from earning my Airline Transport Pilot certificate as soon as I turned twenty-three (the minimum age). I needed this certificate to fly as a captain for a U.S. airline anyway, and I wanted to get it out of the way in case my seniority number started moving up rapidly.

I had passed the written exam and checkride for the Airline Transport Pilot certificate when I was twenty-one, which was shortly after the personnel rep told me about my impending elevation to captain. However, since I didn't meet the age requirement for the certificate at the time of my checkride, the FAA inspector, Theo Moore (my old mentor), issued a "pink slip" to me upon completion of my checkride due to the fact that I didn't meet the age requirement. However, Theo passed me on all of the other checkride tasks. Theo told me to bring the pink slip back when I turned twenty-three. I did and the certificate was issued to me then.

In the summer of 1973, I made an addition to my personal-aircraft fleet by acquiring another Aeronca Champ. However, this time I decided to share its ownership costs and the airplane with two partners. One of them was a certificated pilot and the other was a student pilot. I ended up giving quite a bit of flight instruction in the Champ to my two partners that summer.

In late summer 1973, I flew a trip with a Los Angeles-based captain who had just gotten into the then-new sport of hang-gliding. And he in fact got pretty far in—he was running a hang-gliding school and dealership in the Los Angeles area. He invited me out to LA to learn to fly hang gliders. When our three-day trip terminated in Chicago, I got on an airplane with him and deadheaded out to LAX to find out if hang gliding was really as much fun as he had been telling me about on our trip together.

I spent three days in the LA area learning to fly a **Seagull** hang glider. The Seagull was considered to be the Cadillac of hang gliders and I found it easy and fun to fly. In fact, at the end of my short training period, I flew it (along with an instructor) off a 1,200-foot cliff overlooking **Lake Elsinore**. I was totally hooked on the sport.

I bought a new Seagull from the LAX captain and headed back to Chicago, intending to fly it primarily off the Lake Michigan sand dunes in southwestern Michigan. I took my brother Bob and our mutual friend Vern McGregor along with me on my first foray to the dunes and taught them to fly the Seagull in a down-hill flight regime. I didn't know it at the time, but I was one of the very first hang-glider instructors who held an FAA Flight Instructor Certificate. Almost all of the other hang-glider instructors in the country did not. They were primarily self-taught enthusiasts and ski instructors. At this time, the FAA did not regulate the sport (and they still don't).

About six o'clock in the morning a couple of weeks later, while I was still asleep trying to recover from a particularly challenging three-day trip (the layovers were grueling back then), my brother Bob awoke me with a phone call. He informed me that Wally Phillips, a popular Chicago drive-time host, was talking about hang gliding on his radio show. He was asking if any of his listeners knew anything about the sport. Bob suggested that I give Wally a call.

I rolled out of bed and headed for the coffee pot while tuning into Wally's show on WGN. I knew that Wally hosted the top-rated morning-drive-time radio show in Chicago and that I'd be heard by a lot of people if I called into the show and actually got on the air. At this point in my life, I had not appeared on a radio show although I had briefly appeared on one of the first children's TV shows when I was ten.

I quickly downed a cup of coffee and collected my thoughts. As I poured my second cup, I dialed information to get WGN's phone number. The operator could only find the WGN switchboard's number, so I dialed it believing that at this early hour I would probably get a security guard. That was before voicemail and automated answering systems.

33

I was surprised when Wally's producer picked up the phone. I told him that I was an airline pilot and that I also flew hang gliders. He told me to hold on. Within thirty seconds, Wally came on the line and I was on the air. We talked about hang gliders and flying in general. Wally was a real aviation enthusiast and I talked with him on the air many times over the years after that about flying and spaceflight.

When Wally introduced me at the beginning of that first interview with him, he pronounced my name the way I pronounce it—"cook". However, he didn't clarify that I spell it "Koch". This was an interesting point, because by mid-day my phone was ringing off the hook with callers who wanted to learn how to fly hang gliders. I asked each of the callers how they had gotten my phone number since I had not given it out on-air and it was listed in the phone book as "Koch". The stories of how they found my number were varied and their methods inventive.

I had revealed on-air the community I lived in and that I flew for United. Some knew that "Koch" was sometimes pronounced, "cook" in the Midwest. So, directory assistance could put where I lived and "Koch" together to come up with my phone number. Others found me through United Airlines. The most resourceful called the city hall where I lived and asked if anyone knew me. Since I had lived in the town for several years, many of the folks in city hall knew me and were able to provide the proper spelling of my name. Directory assistance did the rest. By the end of that day, I was convinced that there was a strong market for hang-gliding training in the Chicago area. I received over fifty calls that day and every caller left his or her name and phone number and encouraged me to call them if I decided to start teaching hang gliding to the public.

That evening, I convened a meeting with my brother and Vern. We decided to go into the hang-gliding-training business. This was the second aviation-business start up that I would attempt and my first lesson in how to "bootstrap" a new business into existence through the use of publicity and word-of-mouth marketing.

In the early fall of 1973, my brother Bob, our friend Vern and I started giving skysailing lessons on the weekends. At first, we approached the training on a rather casual basis. We'd schedule a group of five-to-ten students for a morning or afternoon session on Saturdays and Sundays. I would give them a briefing on how to fly a skysail and then we'd work on getting everybody off the ground and back down again without getting hurt. This was how I learned to fly skysails. So, I figured it was a good way to teach skysailing. We flew from a small hill that had been erected in one of the local parks. It was only about fifty-feet high and therefore the flights typically lasted for only a few seconds.

My first several flights in California were from low hills like this one. After my first skysail flight in California, I was surprised at how exhilarating a short flight on one of these wings could be. All of our students flew at least once during their half-day lesson, and I was pleased to see that they were as excited as I had been after my first skysail flight. Skysailing is as close to flying like a bird that I have gotten so far. We also offered organized trips to Warren Dunes State park in southwestern Michigan for students who had completed the basic course. At the Dunes, we were able to get people aloft for much longer flights depending upon how strong the on-shore breeze was.

As far as we were concerned, skysailing in the Midwest was a fair-weather sport. So, we suspended training operations in the late fall of 1973 with a plan to restart training in the spring of 1974. During this first short season, we put over one hundred people through our training programs before we shut down for the winter. They had all come to us through publicity and referrals. We actually made a little money from operations and we were having a ball. And most importantly, nobody got seriously hurt. This convinced us that we had a viable business and I therefore set about writing a formal business plan.

Over the next few weeks, I organized the new business. We called it Apollo Skysailing. I named it Apollo because I had recently flown my Cessna Cardinal down to Cape Canaveral with my brother Bob, Vern and another friend to see the launch of the last **Skylab** mission (November 1973). The astronauts used the **Apollo Command/Service Module** to fly up to Skylab. I was struck by the Apollo name because earlier astronauts had used the Apollo Command/Service Module to fly out to the Moon and back nine times. I didn't like the term "hang glider". So, I used an alternate term that was in limited use in the sport at that time that I liked better—"skysail".

The business planning process forced me to think about the problems our students had experienced in learning to fly a skysail and the hazards to safety that revealed themselves to us in our first few months of operations. I also had to consider whether or not a marketing plan consisting of publicity and word-of-mouth could sustain the business. And I had to define the requirements for whatever courses and programs we were going to offer the next year.

My experiences in that first season made me realize that the typical way hang gliding was being taught was not sufficient if your goal was to train folks who would be qualified to safely fly skysails in the situations we were teaching them to fly in (i.e., low hills, moderately high sand dunes, low winds). I therefore decided to organize the Apollo Skysailing School along the lines of a typical general-aviation flight school.

35

We would offer one-day and two-day training courses that featured ground school in the mornings and group flying in the afternoons. The basic ground schools covered the fundamentals of flight and the particulars of skysailing. In the afternoons, the class would learn and practice the skills needed for the level of skysail flying we were qualifying them for.

We even decided to produce a "simulator" that taught the basics of weight-shift control. It would consist of a **Chandelle** skysail that we planned to purchase as the second skysail in our fleet, and a flight harness just like the one we used for the flight training. We planned to hang the simulator from a sturdy tree limb. The student would step up on a platform below the simulator. This would allow them to attach their harness to the skysail in the same way they would do it when they flew in the afternoon. When the student was securely hooked in, the platform would be removed and the student would be hanging from the skysail just like they would in normal flight.

We would then point the simulator into the wind to give the student the feel of the lift on the wing. Then an instructor would give commands to "turn right", "turn left", "pitch up", "pitch down", "speed up" and "slow down". The student would respond to the instructor's commands by moving the **A-Frame** attached to the skysail in the proper direction. This would, we believed, be effective in teaching the students the "muscle-memory" responses that would make their first flights relatively safe and easy.

In the first season, we identified a significant safety hazard in the basic-training phase. It was the very sudden stop experienced by a student who lost his/her balance on the takeoff run down the hill. They were only moving at a slow-run pace, but if the A-Frame hit the ground, the nose of the skysail would violently pitch down and dig itself into the turf. The resulting sudden stop caused the ill-fated aviator to not only fall down, but to go tail-over-nose through the A-Frame. We had many close calls during our first season, and we were very fortunate that the worst injury was a dislocated shoulder. But I got to the point where I cringed every time a new student started down the hill.

We decided, after some experimentation, to mitigate the risks associated with this hazard by putting lawn-mower wheels on the bottom of the A-Frame and a nose plate on the apex of the skysail. In the event of a failed takeoff attempt caused by the student loosing his/her balance, the A-Frame would just roll down the hill and if the wing pitched down and struck the ground, the nose plate would keep it from digging in and causing a sudden stop. This didn't work on sand dunes. So, we planned to use this risk-mitigation technique only in the basic-training phase where we flew off turf-covered hills.

A more serious hazard that we identified, however, was the nature of the existing hang-gliding culture. Since over ninety percent of our clientele were young men in their twenties, a macho attitude was the norm and " testosterone dementia" freely coursed through the culture. During that first season, we had noticed that the safety hazards associated with this type of culture could be of a scale that would make it a bad idea to continue in the business of teaching people to fly skysails.

To mitigate the risks associated with this hazard, we decided to create a "real pilot" oriented culture that was more like that found in general-aviation flying clubs. There would still be a lot of fun, competition and adrenaline-producing thrills, but we would focus on acquiring the basic knowledge and skills needed to fly skysails safely while encouraging everyone to assess the risks associated with a particular flight before attempting a takeoff. I spent almost all of my free time during the winter of 1973 – 1974 working on completing the Apollo Skysailing School business plan and getting things ready to relaunch flight-training operations. And as it turned out, I had plenty of free time.

In October 1973, the first oil crisis reared its ugly head. Almost immediately, United Airlines cut its schedule and as a result had a surplus of pilots on the payroll. My relative-seniority position rapidly eroded. It wasn't long before I had been "surplused" off the DC-8 and sent back to flying as a 727 second officer.

However, there was a silver lining to this cloud. As a fairly senior 727 second officer, I could bid a reserve schedule. Since there were now plenty of reserve 727-second officers in Chicago and the demand for them very low due to the schedule cut backs, I was rarely called out to fly. In fact, during the protracted slowdown caused by the oil embargo I only flew once or twice per month.

This turned out to be a lucky break because I could focus almost entirely on getting the skysailing business ready for the spring-1974 restart of operations. Actually, it was every entrepreneur's dream. I was drawing full pay from United, but I had all the time I needed for the new business. And I was able to keep my flying for the Army Guard down to one weekend per month. The "get-ready-for-deployment" order our Guard unit had received was now only a dim memory. Official military involvement in the Vietnam War had ended in the summer of 1973 and we were told to stand down shortly thereafter. The routine in the unit returned to a pre-Vietnam War pace with weekend drills once a month and a two-week summer camp. I reduced my flying with the Guard to about fifty-hours a year.

I flew an eclectic mix of aircraft in the Guard. My unit still flew Korean War-era **Hiller OH-13 Raven** light-observation helicopters, **Cessna L-19/O-1 Bird Dog** light observation/liaison fixed-wing aircraft, **de Havilland DHC-2**

Beaver fixed-wing utility aircraft and **Sikorsky CH-34 Choctaw** medium-transport helicopters. And of course we flew the ubiquitous Bell UH-1 Iroquois (Huey) assault/attack helicopters and Bell OH-58 Kiowa observation/utility helicopters. I enjoyed flying them all and built up over 750 hours of helicopter time while serving with the Guard and Reserves.

During my service with the Guard and Reserves, I was steeped in the Airmanship 1.0 culture that military aviators lived in at that time. This culture was, of course, highly mission oriented, macho and high-risk. Our hazard-recognition and risk-mitigation techniques were woefully inadequate. And the human-factors aspects of airmanship were routinely ignored. Nevertheless, I was fortunate to have survived the experience with so many fond memories and so many lessons learned.

In the spring of 1974, we began season two of Apollo Skysailing. Over the previous winter, I had purchased a new Chandelle hang glider for use as our "simulator" and as an addition to our training fleet. I also bought another Seagull skysail to augment the training fleet. We opened a store in a strip mall near Roselle Airport and hung one of the skysails from the ceiling in the store during the week. The store enjoyed brisk foot traffic since it was located next door to a saloon that attracted a lot of our potential clientele.

A significant number of new students came to us through the publicity campaign we launched with the restart of operations in 1974. We garnered a lot of free "ink" from the local press. It wasn't long before my brother Bob had acquired the nickname "Sky Bob" because of his coverage in a *Chicago Sun-Times* article. We gave Bob a **"Sky Bob"** plaque with clippings from the article to immortalize his new name. We still call him that today.

Vern McGregor was also anointed with a nickname. It was "Crash McGregor". He earned this sobriquet by crashing more times than any of us. We should have anticipated Vern's crash record because while serving as a member of the deck crew on an aircraft carrier early in the Vietnam War, he narrowly escaped death three times. One of those episodes included being blown off the carrier's deck by a maneuvering aircraft.

During our first season, we had become dissatisfied with the hill we were using for basic-skysailing training. Over the winter, I was able to secure permission from the **Norge Ski Club** in Fox River Grove, Illinois to use their ski-jump hill for our skysailing-training operations. They were more than happy to let us use their hill because the actual ski-jump structure that stood atop the hill had recently burned down and the club needed funds to rebuild it. The hill below the ski jump was about one-hundred-feet high and the bowl below the hill provided a good landing area for our skysails.

Chapter 1: My Pursuit of More Enjoyable Personal Flying

Our Apollo Skysailing crew grew from my brother Bob, Vern and me to about fifteen people who staffed the store seven-days a week, taught skysailing, served as ground-support crew and did whatever else it took to get things done. We held ground-school classes in the store in the mornings on Saturday and Sunday, and flew at the Norge Ski Club in the afternoons. There were social gatherings at the store every Saturday night and instructors and advanced students made regular trips to Warren Dunes on the weekends. In addition to selling skysailing lessons, we sold Seagull and Chandelle skysails to our clients who had qualified to fly them in our training programs.

We created an airmanship culture by weaving basic airmanship into our training programs, publishing a monthly newsletter that featured airmanship issues related to flying skysails and through the use of the then state-of-the-art 8mm-home-movie technology. Our crew was tightly organized, standardized and trained to model the behaviors we based our airmanship culture on.

During our second skysailing season, we put over five hundred people through one or more of our programs and we sold a couple dozen skysails. The innovations we had introduced to skysailing training and marketing proved to work approximately as we expected. In fact, the wheels and nose plates on the skysails helped us to get through this second season with no injuries. The business once again made some money and we had enough in the coffers to keep the store open throughout the winter season when we didn't plan to do any skysailing. In November of 1974, we closed down skysailing-training operations for the winter. I was more convinced than ever that Apollo Skysails was a viable business and that it had the potential to introduce thousands of people to the world of aviation. My thoughts that winter turned to expanding the business.

By this time, my relative-seniority position at United had once again been adjusted downward due to a recently instituted round of layoffs. This reduction in personnel was caused by the continued disruption and escalating costs associated with the 1973 oil embargo. By the winter of 1974, my seniority number would only buy me a slot as a **Boeing 737-200** second officer. This was my last stop before being furloughed if United decided they needed to further reduce the size of the pilot corps.

The Boeing 737 second officer issue had already become a bone of contention among the airlines and the pilot groups that made up the Air Line Pilots Association (ALPA). The 737 is one of the first jet airliners to be certified for a two-pilot crew, but United management had decided to fly it with three crewmembers for various reasons related to safety and professional development. They believed that it was better to have three sets of eyeballs in the cockpit rather than two. And most of us agreed that the second officer position was the best place to learn how to be an airline pilot. However, United's rivals who flew **DC-9s** (a competitive airplane) crewed them with two pilots.

At this point, the resolution of this disparity was not considered to be of much importance. At the time, the Civil Aeronautics Board (CAB) tightly regulated the airline industry. The CAB set all of the fare levels across the industry. The CAB also regulated the routes and types of equipment an airline flew. The fares were always set at a level that allowed an airline to recover its costs and make a reasonable profit. So, it wasn't very difficult for United to cover the costs of its three-pilot 737 crews. This was before the era of airline deregulation and the resulting cutthroat competition between the airlines. However, as we shall shortly see, this crew-complement issue was to raise its ugly head by the end of the decade. The pursuit of a resolution to the issue set me on a new path in the development of my approach to airmanship excellence — more about that later.

For the next couple of years, I enjoyed a spot very near the top of the 737-second-officer list in Chicago. This was due to my luck of being one of the one hundred junior pilots that United planned to furlough. I was lucky because the United pilots agreed to a pay freeze and some changes in their training options while United was going through tough times. The quid pro quo for the pilots was a no-furlough clause in our agreement with the airline.

Since my seniority put me at the "top of the bottom" of the seniority list, I ended up as one of the top three or four 737-second officers in Chicago. This allowed me to bid reserve every month. Since there were about one hundred reserve 737 second officers in Chicago, and only about 30 were needed to cover the reduced schedule, I was rarely called out to fly a trip. In fact, I would usually volunteer to fly a couple of times a month just to keep my flight-crew proficiency up to snuff.

This fortuitous situation once again allowed me to draw full, if somewhat reduced, pay from United while enjoying a bounty of free time to concentrate on Apollo Skysailing's third season. Throughout the winter of 1974 - 1975, I put in a lot of time on the revised plan for the business. The business plan we implemented in Apollo Skysailing's prior season vetted the marketing, training and operations plans that we had created during the winter of 1973 - 1974. The season also provided us with some lessons learned and some opportunities to be considered.

The most important lesson that we learned was that for the basic-skysailing training to be optimally effective, we needed to create a way to get a skysail from the bottom of the hill back up to the takeoff point without having to carry it. Even the young bucks that made up the vast majority of our market regularly complained about being rather quickly fatigued by that task. And it took too long for a complete takeoff-flight-landing-reposition cycle. We recognized that we needed to solve this problem before we could expand our customer base to a point where the business would be really interesting.

The most important safety lesson we learned was that we needed to find a way to keep neophyte skysailers on the straight-and-narrow as they flew down the hill until they got the (pardon the pun) hang of things. Although we tried to build the "muscle memory" a skysailer needed to be able to control a skysail through the use of our home-grown simulator, we found that there were some students who just needed to fly down the hill several times until that muscle memory was strong enough to allow them to control the skysail without having to consciously think about it. Unfortunately, we had some close calls when a student would stray too far off course and head for an obstacle off to the side of the desired flight path.

We proposed to solve both of those problems by designing a cable system that was supported at the top and bottom of the hill by telephone poles. We planned to affix a three-foot-diameter ring on top of the skysail. The cable would pass through the ring so that the skysail could veer only slightly off course before it would reach a limit imposed by the cable-ring system. A second cable would be attached to the aft end of the skysail and a high-speed electric winch mounted on the pole at the top of the hill would pull the skysail back to the takeoff point after a flight down the hill.

We made other refinements to the design, but the basic idea was to use the electric winch to reposition the skysail from the landing point to the takeoff point without the student having to detach from the skysail. In other words, they'd go along for the ride back up hill rather like a skier. The cable-and-ring would prevent the student from injury by getting off course and colliding with a hazard. We would therefore enhance the safety of our training, make it more effective and broaden our market by making it much easier physically to learn to fly skysails.

Up to this point our marketing plan consisted solely of publicity (which we got a lot of) and referrals from our customers. Going into our third season, I decided to rely on this marketing system since it was doing so well for the business. In fact, I decided to open a second store in Glenview, Illinois to get closer to another geographic market.

During our second season, we had bought and sold a few skysails. However, there wasn't much of a margin in it. My brother Bob reverse engineered a couple of the models and then determined what it cost to build a high-quality skysail with all the safety enhancements that were available at the time or that we could invent. The results were interesting. It appeared to us that producing and retailing our own skysails to a big-enough market could yield an attractive return on investment.

After significant additional research, we decided to produce our own line of skysails. We started out with a standard model and planned to add other models as the customer base grew. By the time we restarted training operations in the spring of 1975, we had set up a skysail-production line in an old shoe factory in Elgin, Illinois, and we had produced five new Apollo Skysails—three for use in our school and one for display in each of the two stores. We had the Glenview store open by this time.

Apollo Skysailing's third season started strong. The marketing efforts in both stores were working well and we got a lot of attention from the TV stations, radio stations and newspapers in the Chicago area. I even wrote a short article about skysailing that was published in a local magazine.

Our skysailing-training classes were booked up a few weeks in advance. We introduced new training films that we had produced in-house into the skysailing-training programs and opened up more activities for our customers to experience the airmanship culture we were building. Throughout the season, we tested a prototype of our cable-and-ring system for getting skysails back up the hill and keeping students on course. Our skysail factory turned out several more skysails for use in our training program and for sale to our customers.

By the end of this third season in the fall of 1975, over one thousand people had gone through one or more of our skysailing-training programs. We had also established the name Apollo Skysailing as the gold standard for safe, effective skysailing training. And our skysails were popular with our customers. Apollo Skysailing had proven to be successful by most measures. We had made a profit in each of the first three years of operations, our customer base was expanding, our skysails were a success in the marketplace and I had a pretty clear vision of where I wanted to take the company.

As the winter of 1975 - 1976 set in, I went back to the drawing board and started working on a business plan to realize that vision. By mid-winter, I had come to the conclusion that I wanted to continue in the skysailing business only if I could gather the resources I needed to implement the business plan I had in mind. I believe this attitude resulted from my entrepreneurial spirit that always drives me to the next step rather than just maintaining the status quo.

The key resource we needed to implement the new business plan was capital. I was twenty-nine at the time and although my United pay had provided enough discretionary funding to allow me to get Apollo Skysailing to this point, I knew from my analysis of the business plan that I would have to bring in outside capital to make it work. I recruited a couple of my skysailing students who had a background in the financial industry to assist me in seeking

Chapter 1: My Pursuit of More Enjoyable Personal Flying

the capital we needed. We all did a lot of work over a very short period of time (oh, the energy of youth) and we started talking with potential investors before the winter was over. The initial reactions to our overtures were not encouraging. As I look back on that experience with the benefit of 20/20 hindsight, I can clearly see that our plan was not sophisticated enough, the market was too unproven and we were all way too young to attract the kind of capital we thought we needed to go forward.

As we approached the decision point for the launch of a fourth Apollo Skysailing season, I was seriously considering walking away from the business because I could foresee only relatively slow growth for it if we continued on our current trajectory. I was more interested in breaking new ground than putting my efforts into managing a business that wasn't fulfilling my expectations.

I struggled with this dilemma through the "dog days of winter" (February and March) in 1976, feeling the pressure mount as we approached the go/no-go decision point for re-launching flight-training operations. As I recall, it was sometime around the end of March when I read an article in Airline Pilot Magazine that jumped out and grabbed me by my collar. That article, the long-drawn-out end of a cold winter and the pressures on me at the time, caused me to make an immediate decision to leave town right away.

My destination was the **Johnson Space Center** (JSC) near Houston, Texas. The Airline Pilot Magazine article was about an announcement by NASA that they were recruiting a new class of Space Shuttle astronauts. Normally, this news would have been of only mild interest to me. I had always been a science-fiction fan and I was interested in spaceflight from a professional perspective. I really believed at that time that the pace of development in manned spaceflight would inevitably lead to commercial spaceliners that would be flown by airline pilots like me. However, I thought that my opportunity to fly spacecraft would probably arrive near the end of my airline-pilot career—still over three decades into the future.

What really motivated me to jump in my car (I had by then moved up from the Volkswagen Beatle to a Cadillac Fleetwood Brougham.) was the news in the article that NASA was, for the first time, accepting applications for their open pilot-astronaut positions from professional pilots who were not necessarily military test pilots. In fact, as I recall, the requirements only specified flight experience in jet aircraft and a technical degree. When I saw that news in the article, my imagination immediately started working. At that time, the goal of becoming a NASA astronaut seemed less than attainable, but I couldn't resist the temptation to find out if I had any chance at all. And, the facts that I badly needed a change of scenery and it was warm in Houston served to give my quest at least some legitimacy.

My drive to Houston from Chicago provided me with ample time to let my imagination work on the possibilities. By the time I reached my hotel near JSC, I had somehow convinced myself that I was just the kind of young pilot NASA was looking for. I arrived at JSC all bright-eyed and bushy-tailed about mid-morning the next day and immediately made a beeline to the NASA personnel office. At that time, JSC was open to the public and I had no difficulty just walking into the office.

It wasn't long before I was face-to-face with the bureaucracy I had become used to in the Army and working for the largest airline on the planet. The personnel clerk who fielded my inquiry did not immediately set off the fireworks and usher me into an interview for the pilot-astronaut job I was interested in. Instead, she handed me a one-quarter-inch-thick stack of civil-service application forms and told me to mail them in when I had completed them.

Since I had traveled to JSC to find out more about the job, the selection criteria and the selection process, I asked the clerk for more information on these subjects. My request was answered with an 8.5" x 11", two-sided, one-color, three-fold brochure. Although I applied my best charm attack on the clerk, this was as far as I got. So, I retreated to my hotel to go over the brochure and the application forms.

The information contained in the brochure was essentially the same as that in the Airline Pilot Magazine article. And the application material didn't shed a whole lot more light on things. After going through everything, I was very discouraged about actually getting even close to a pilot-astronaut job. I decided to repair to the hotel restaurant for dinner and a couple of drinks to think things over.

After a very good seafood dinner and an after-dinner drink or two, I had a sudden inspiration about how I could find out more about becoming a NASA pilot-astronaut. I decided to go into the JSC Public Affairs Office the next morning and tell them that I was a published "aerospace journalist" (Remember, I had an article about skysailing published a little over a year earlier.) and airline pilot and I wanted to write an article for *Airline Pilot Magazine* about how one became a NASA pilot-astronaut.

The next morning, after talking with the editor of *Airline Pilot Magazine* on the phone and securing an invitation to submit my proposed article to the magazine, I presented myself at the JSC Public Affairs Office. This time, my story resonated with the clerk who took my inquiry and she introduced me to a NASA public-affairs officer—Milton Reim. Milt and I immediately hit it off.

He told me that my media credentials were good enough to get me the grand tour of JSC and maybe even interviews with a few current astronauts. Milt personally showed me around JSC and introduced me to several Space Shuttle managers, engineers and technicians. He even arranged for me to fly the Space Shuttle simulator and to sit in on a ground-school class that was being held for the current crop of shuttle astronauts.

After a lunch break in the JSC cafeteria (I, of course, bought Milt lunch.), Milt suggested that I interview the two most-senior astronauts at JSC—John Young and Deke Slayton. Young was the chief of the astronaut office and Slayton was NASA's director of flight-crew operations. Young had flown in the Gemini and Apollo programs and he was slated to be the commander of the first Space Shuttle mission (which eventually flew in 1981). Slayton was one of the original Mercury 7 astronauts and he had flown on the Apollo-Soyuz test project in 1975. Needless to say, I jumped at the chance to meet these two icons of American spaceflight.

Milt ushered me into Young's office and introduced me to him and Slayton. However, before I could begin my interview, they both peppered me with questions about being an airline pilot. I ran into this phenomenon frequently as I met NASA personnel at JSC. It seemed that the NASA folks, including the astronauts, considered an airline pilot to be a minor celebrity (as did most of society in general at that time). As I was to learn later, most JSC personnel were so used to talking with astronauts that they no longer considered them to be anything special. A real, live airline pilot on the other hand was a rarity at JSC.

I asked Slayton and Young about the possibility of the Space Shuttle becoming a commercial operation once the Orbital Flight Test period was completed. They both stuck to the then-current party line that of course the Space Shuttle would one day, in the not too distant future, be carrying commercial payloads to low-earth orbit. My follow-up question was, "Do you think airline pilots will be flying the Space Shuttle when commercial operations begin?" I now realize that this question must have appeared to be a bit naïve to these seasoned astronauts, but I was asking it as a logical follow-up to their comments about commercial Space Shuttle operations.

I was somewhat shocked, and I sensed Milt was also, when both Young and Slayton started laughing. When they calmed down, they told me in no uncertain terms that airline pilots were not capable of learning to fly the Shuttle. It was their firm belief that only test pilots could fly such a complex machine. After that rather humiliating response, I thanked both of them for the interview and asked Milt if he could escort me to the next stop on our tour.

45

As we walked over to Mission Control, Milt shared his viewpoint on the arrogance of astronauts in general and reaffirmed for me NASA's stated intention to turn the Space Shuttle over to the private sector for commercial operations once flight-testing was completed. He told me that although he was certainly not qualified to comment on Young's and Slayton's position on whether or not airline pilots could learn to fly the Space Shuttle, he was more than willing to make any resources available to me that I might need to prove them wrong.

Late in the afternoon, after my tour was completed, I went back to my hotel to clean up my notes and think about what I had seen and heard during one of the most interesting days I had ever spent. I had talked with several very knowledgeable spaceflight experts, sat in on a Shuttle-astronaut ground school and interviewed two of America's most-revered astronauts. And I had flown the Space Shuttle simulator!

It appeared to me that learning to fly the Shuttle was much the same process as I had gone through in my initial airline-pilot training. I had found nothing that pointed to it being beyond of the capabilities of a typical airline pilot to learn spaceflight operations. Young and Slayton had presented me with the challenge of a lifetime. And Milt Reim had offered to help me address that challenge. Based on these findings and Milt's offer to provide me with the access I would need to prove that an airline pilot could learn to fly the Space Shuttle, I decided to take up the challenge. I spent the rest of the evening composing my article for Airline Pilot Magazine and planning my approach to addressing that challenge.

The next morning, I met with Milt in his office and outlined what I wanted to do. My plan was to get the article published and then return to JSC to begin working on a book about learning to the fly the Space Shuttle. Milt told me that with the publication of the article, I would be able to secure a NASA media badge that would essentially give me free run of JSC. He also reiterated his offer of assistance in my quest to prove that an airline pilot could learn to fly the Space Shuttle.

Needless to say, I was on Cloud Nine on my drive back to Chicago. By the time I returned home, I had pretty much convinced myself that I could learn to fly the Space Shuttle. I would be more than happy to prove Young and Slayton wrong. Over the next couple of weeks, I polished and submitted my article to *Airline Pilot Magazine* and I poured over the Space Shuttle manuals that Milt had supplied me with. My article was published in *Airline Pilot Magazine* shortly after I submitted it.

I continued my research into all things related to spaceflight and mulled over my options. My situation at United was, at least for the time being, stable and I could look forward to at least several more months of full pay with a very light workload. So, I had the foundation to either launch another season of Apollo Skysailing or to go for the challenge that Young and Slayton had unwittingly presented me with. After consultations with my brother and a whole lot of soul searching, the siren call of spaceflight won out over the challenge of taking Apollo Skysailing to the next level.

The skysailing business had taught me a lot about airmanship on the most basic level, and it provided me with the opportunity to develop airmanship-related skills like training-program and airmanship-culture development. But I realized that the spaceflight challenge would put me on an airmanship-development path that only a relatively few airmen had experienced. I couldn't pass up this opportunity of a lifetime.

Shortly after making my decision and the publication of my article in *Airline Pilot Magazine*, I established a base of operations near JSC. I rented a cottage on Galveston Bay in Kemah, Texas. I set up a personal office that included one of the first PCs. I used it primarily for word processing. Milt Reim was good to his word and I soon had my very own JSC media badge. I set to work in earnest to learn as much about spaceflight as I possibly could.

Over the next three years (1976 – 1979), I established myself in the JSC community by serving as the chair of the publicity committee for the JSC chapter of the **American Institute of Aeronautics and Astronautics** (AIAA). I also became the founding president of the local chapter of the Aviation/Space Writers Association and the founding director of the first "Space Week" celebration to be held in Houston. I'm happy to report that Space Week is still an annual event that is celebrated across the U.S. I also served on a NASA-sponsored national committee that was working to gain public support for the Space Shuttle program.

At this time, I was also enrolled in the **Embry-Riddle Aeronautical University's** Bachelor of Science in Aerospace Studies program. I needed to complete a degree to qualify for selection as a NASA astronaut and I hadn't taken the time to get a degree prior to this time. I had briefly attended Elgin Community College in 1965, but had more or less dropped out of school because of my heavy flying schedule as a flight instructor. It seemed that every time I was offered the choice of flying an airplane or sitting in a classroom, I chose the flying. I, and my mentors, believed that it was more important to my sought-after airline-pilot career to build my flying time than to complete a bachelor's degree. However, they counseled me that I would have to get a degree to be hired by an airline (as was the norm at that time). At that point, I didn't think I would be hired by an airline until I was in my thirties based on what was then happening in the airline industry.

So, my strategy was to rapidly build my airmanship to the point where I could get a relatively good-paying, flextime job flying for a corporation or a charter operator and then go back to school on a part-time basis to complete my degree. This strategy would have put me in a great position by the time I turned thirty with thousands of hours of flight time and a bachelor's degree. This was the typical winning combination for securing a coveted airline-pilot job at that time.

Luckily, I didn't have to fully implement this strategy. When United told me in 1967 that they were going to hire me the following year when I turned twenty-one, they also told me that they didn't care if I returned to college or not. Since I didn't have an immediate need for a degree, I put that task on my "someday" list and continued to do what I loved best—flying airplanes—chalking it all up to my good fortune in not having to spend countless hours in a dreary classroom.

I later discovered that United was running a test program in the mid-1960s that I was unknowingly a part of, and that's why I didn't need a college degree to get hired. At that time, the airlines were finding it increasingly difficult to find suitable new-hire pilots with the several-thousands of flight hours that they preferred. In fact, in the mid-1960s, several of my students were hired by United with no flight time at all. They were given a year to earn their professional certificates and then they were trained by United and put on the line just like any other new-hire pilot.

United's multi-year study was designed to find out what the correlation was between flight experience, age and education level at time of hire and later success as an airline captain. Therefore, United made it a point to hire several new pilots with low flight hours and only a high-school diploma. Although I had about two thousand hours when United hired me, I had only a few college credits on my transcript. I became aware of the study when I was working with Capt. J.D. Smith, United's Senior VP for Flight Safety and Industry Affairs, in the early 1980s. J.D. told me that United had discovered that formal education level didn't really make any difference in producing a high-quality airline captain.

By 1978, my relative seniority at United was finally heading in the right direction and I was able to win a slot as a Boeing 737 first officer at United's Newark base. This was the first copilot bid my seniority number would buy me. I was thirty-one years old. I went through 737 first officer training almost exactly ten years after I had completed my new-hire training. The United personnel rep's prediction about me getting a captain bid before I turned twenty three was obviously a little off.

Chapter 1: My Pursuit of More Enjoyable Personal Flying

Needless to say, the prior ten years had been very frustrating for a young, ambitious airman because of slow progression through the ranks. Instead of moving from second officer to captain in two-to-three years (as had been predicted when I was hired), it took me ten years just to move up to first officer (copilot). And when looked into the future to predict when I might make captain, it appeared that it would take at least as long as it took me to make copilot. This professional frustration only added fuel to my desire to become a NASA astronaut. I was fully prepared at this time to resign from United if NASA offered me a job flying the Space Shuttle even though NASA-astronaut pay was less than half of what I was by then making at United.

After I completed my 737-first-officer training in late 1978, I refocused on the research for my book on learning to fly the Space Shuttle. I had my own complete set of Space Shuttle manuals—just like the NASA astronauts. I regularly attended basic-spaceflight classes at JSC along with the crop of new NASA astronauts, and I had access to any NASA or contractor expert I cared to interview. I even flew several more hours in the Space Shuttle simulator flying the same scenarios the NASA astronauts were using. I could count among my friends and associates several current and former astronauts and many of the pioneers of manned spaceflight. I was also securely embedded in the NASA culture at JSC and at the Kennedy Space Center in Florida, and I regularly met with top NASA officials in Washington.

In 1978, I started a small public-relations/corporate-communications consulting business that I called Aerospace Communications (AeroComm). It was based in Clear Lake, Texas across the street from JSC. In addition to a small support staff, I had several independent contractors working for me who were very experienced, well-liked aerospace PR and corporate-communications folks. We participated in several aerospace-related public-relations campaigns.

By 1979, I was convinced that a typical airline pilot could learn to fly the Space Shuttle in commercial operations. I had also positioned myself to at least have a shot at a NASA pilot-astronaut slot. As I was politicking to be considered for the next group of NASA Shuttle astronauts to be hired, I was presented with another challenge that I couldn't resist.

I was sitting at the head table at a dinner meeting of the Houston chapter of the American Institute of Aeronautics and Astronautics. I was at the head table because as the publicity committee chair, I had to give a report to the assembled members. Sitting next to me was **Dr. Robert Gilruth**, an aerospace scientist, engineer, and a pioneer of the American space program during the glory days of Mercury, Gemini and Apollo.

During his forty-year career with NASA and its predecessor, the National Advisory Committee for Aeronautics (NACA), Dr. Gilruth led many of the nation's leading-edge flight-research and human-spaceflight operations. He had been the director of the NASA Manned Spacecraft Center (later renamed Johnson Space Center) from 1961 until 1971. By 1979, Dr. Gilruth had retired from NASA, but he still served as a consultant to NASA and the aerospace industry. He was at the meeting to accept an honor we were giving him for his pioneering service in the realm of manned spaceflight.

I had met and talked with Dr. Gilruth several times at JSC and at various spaceflight-related meetings that I had attended across the country. He had always encouraged me to pursue my dream of flying the Space Shuttle commercially. However, on this particular evening, he went way beyond merely encouraging me.

After our second martini, Dr. Gilruth leaned over and casually asked me if I thought United Airlines would be interested in taking over commercial Space Shuttle operations when the Shuttle's orbital-flight testing was completed in 1980 (The first Shuttle flight later slipped to April 1981.). At first, I thought he was just engaging in friendly conversation, but it soon became clear that he was asking me this question on behalf of the NASA administrator, Dr. Thomas Paine. I told Dr. Gilruth that I didn't know the answer to his question, but I would try to get one from United's management.

Unfortunately, at that time I didn't have access to anyone at United who might be able to answer Dr. Gilruth's question. My world at United was centered on the pilot community. Virtually everyone that I knew well at United was a pilot, a flight attendant or someone who worked in flight operations. I had not been at all active in pilot or company politics. I would fly my trips and then go home and forget about United Airlines. And my relatively young age (I was then thirty-two.) and low seniority afforded me very little status within the United pilot community.

I was based back in Chicago now and my United Flight Manager at ORD was Dick Roundtree. I decided to talk with Dick about this incredible opportunity that had been entrusted to me by Dr. Gilruth. Luckily, Dick was only a few years older than me, he had an open, creative mind and he was hoping to move up through the management ranks at United.

Most of the pilots at United had read my article about the Space Shuttle in *Airline Pilot Magazine*. So, Dick was somewhat familiar with my interest in commercial space operations. However, relatively few people at United, none of them in management, knew what I had been doing at JSC for the past three years. I filled Dick in and told him about the door that had been opened to possibly the best deal United was ever going to see. Namely, to take over all commercial Space Shuttle operations with the support of a NASA cost-plus contract.

Chapter 1: My Pursuit of More Enjoyable Personal Flying

Dick immediately saw the implications of what I told him. He assured me that he would try to set up a meeting for me to tell my story to United's brand-new CEO—Richard J. Ferris. To my surprise, he was successful. I hastily put together a 35-mm slide presentation and hurried over to Ferris' office in the United Airlines Executive Offices in Elk Grove Village, Illinois.

However, my meeting with Ferris only lasted about five minutes. He told me he wasn't interested in my slide presentation and he thought I was off my rocker concerning United taking over Space Shuttle operations. I was unceremoniously ushered out of his office with an admonition to forget about spaceflight and to instead concentrate on flying airplanes for United.

I reported back to Dick Roundtree on my meeting with Ferris and told him that I was seriously considering resigning from United to concentrate on my pursuit of a NASA-astronaut slot. Dick tried to discourage me from making such a radical decision when I had a long and rewarding airline-pilot career ahead of me. He offered to set up a meeting for me with the newly elected chairman of the Air Line Pilots Association (ALPA) United Master Executive Council (UAL-MEC) John Ferg.

Ferg was no more receptive to my ideas than Ferris was, but he at least granted me the time to make my slide presentation. After the meeting with Ferg and the lukewarm reception I had received for my ideas, I told Roundtree that I was definitely going to resign from United as soon as I got my ducks in a row. However, a couple of weeks after my meeting with Ferg, Ferg called me at home and told me to get on an airplane to LAX. There was a UAL-MEC meeting in progress there and he wanted to talk with me about an important deal. He wouldn't tell me any more about the deal, so I got on the airplane to LAX to find out what was going on.

When I met with Ferg in LA, I was offered a deal that I couldn't refuse. If I would contribute my communications expertise to solve a pilot-culture-change problem for United Airlines and the UAL-MEC, I would be installed as a special assistant to Percy Wood, who was the President and COO of United, to look into the Space Shuttle opportunity. I would also be authorized to establish a UAL-MEC-level committee that would be designed to get United's pilots interested in the idea of commercial spaceflight. I later named it the UAL-MEC Professional Outlook Committee. This was in 1979 on the eve of airline deregulation.

The pilot-culture-change problem revolved around a resolution to the crew-complement issue that would become a big thorn in the airline industry' side after deregulation of the industry took effect. United and some other airlines were flying Boeing 737s with three crewmembers while its competitors flew their Douglas DC-9s with two crewmembers. In the newly competitive world of airline deregulation, United wanted a level playing field and ALPA wanted to put to rest any intra-association rivalries among its member airline-pilot groups.

After some very interesting ALPA political maneuvering, the association at the national level decided to launch an effort to get all of the ALPA-member pilot groups to agree to walk off the job for three days to force a resolution to the crew-complement issue—one way or the other. It was my job to find a way to convince all of these very individualistic pilots, who thought of themselves more as management types than labor types, to walk off their jobs without authorization from their companies and take up positions on picket lines at every major airport in the U.S. If we could pull that off, virtually all airline traffic in the country would be shut down since around ninety percent of all airline pilots in the U.S. belonged to ALPA at that time.

When John Ferg told me about this part of the deal, I at first thought it was going to be very difficult to make it happen, especially in the ninety-day timeframe I was given. I asked John why he thought I could accomplish the mission since I had absolutely no experience in ALPA politics. He told me that the reason he and Ferris were asking me to tackle the problem was because they felt that I had the requisite communications skills and insight into the current airline-pilot-culture zeitgeist. I wasn't sure they were right, but I was certainly up for the challenge.

Towards the end of 1979, I was sent to ALPA headquarters in Washington, DC to head up ALPA's campaign to convince its members to shut the U.S. airline system down for three days to force a resolution to the crew-complement issue. I dubbed the project "Operation USA". Within the allotted ninety days, eighty-five percent of all ALPA pilots had voted to support the walk out. Shortly thereafter, President Reagan appointed a presidential commission to decide the issue, and they did. Unfortunately, from my point of view, they decided on two crewmembers instead of three on all two-crewmember-certificated airliners. Ferris and Ferg were good to their words and I was assigned to work with Percy Wood on the Space Shuttle opportunity. I was also installed as the chair of the UAL-MEC Professional Outlook Committee.

I thoroughly enjoyed working with and getting to know Percy Wood. My job as his special assistant was to bring him up to speed on the Space Shuttle opportunity, introduce him to my contacts in NASA and the aerospace industry and to continue to monitor the Shuttle landscape for additional insight into the deal. Percy and I traveled together quite a bit on Space Shuttle business. So, we got to know each other pretty well. Occasionally, other United Airlines experts would join us on these fact-finding missions.

Chapter 1: My Pursuit of More Enjoyable Personal Flying

1980s

During 1980, I spent most of my time at United working on the Space Shuttle deal. My special-assignment status allowed me to displace any Boeing 737 first officer in Chicago from his assigned trip, or any part of a trip, so that I could maintain my airmanship proficiency. I was also authorized to pick up any open flying on the schedule for the same reason. I frequently displaced my fellow copilots when I was enroute to or from some Space Shuttle business. As I recall, I managed to fly for United around twenty hours a month during this period.

In late 1980, I was asked to design and implement another pilot-culture-change program, this time exclusively for United Airlines and its pilots. The challenge this time was to convince the United pilots to agree to the first concessionary labor agreement at a major airline following deregulation. A lot was at stake on both sides of this issue, but the main goal was to change the thinking of United's pilots from "regulated-industry" to a "deregulation" point of view. In 1981, after another ninety-day campaign that we dubbed "Operation Blue Skies", eighty-five percent of United's pilots voted to accept the new concessionary contract.

In my design of Operation Blue Skies, I expanded on the culture-change methods I had developed in Operation USA a year earlier. I was the first communicator to use the then-new personal-video technology on a wide scale in the airline industry. I convinced United to put VHS video players and large-screen (for the day) TVs in all of the pilot lounges in United's pilot bases. I then produced a series of video productions that highlighted the key issues under discussion and distributed the videos to the pilot bases. Line pilots viewed the videos as they were passing through the pilot lounges. Feedback from these video productions was collected by the existing ALPA political structure and passed along to the leadership. We then modified our messages as necessary and produced and distributed new videotapes to address the feedback.

I was also the first to employ a recently available technology, video teleconferencing, in the airline industry. We held an Operation Blue Skies "road show" in Denver that several hundred United pilots attended. We produced the meeting as a TV show and uplinked the program to a satellite. The program was then downlinked to large meeting halls in United's pilot bases where a large number of United pilots were in attendance. We also recorded the program and distributed it in segments for use in the VHS players in the pilot lounges.

In mid-1980, my military flying career came to an end. A couple of years earlier, the National Guard decided to move the helicopter company I had been a part of since 1967 to Decatur, Illinois. The move made it all but impossible for me to continue to serve in that unit. So, I transferred to a U.S. Army Reserve helicopter outfit at DuPage Airport in West Chicago, Illinois. About a year after I joined that unit, it was moved to Glenview Naval Air Station in Glenview, Illinois.

It was getting progressively more difficult to fit my Army flying into my United Airlines flight schedule, and my extracurricular professional activities were taking up most of my time when I wasn't flying for United or working on the Space Shuttle deal. Also, I was expecting my first child at the time. So, after about thirteen years in the Army Guard and Reserves, I reluctantly made the decision to resign from military service.

In late 1980, I was asked to consult to the United-pilot-training team that was getting ready to launch the most-important pilot-culture-change attempt in the history of aviation. Their assigned task was to take the first big step away from an Airmanship 1.0 culture to the emerging Airmanship 2.0 culture by introducing a new required training course. This course was called "Command/Leadership/Resource Management" (C/L/R).

The training had been mandated by the FAA in response to a string of high-profile airline crashes. The last link in the chain of events that led up to this requirement was the December 1978 crash of a **United Flight 173** at Portland, Oregon. United had to produce, prior to the 1981 deadline, a program to deal with the Human Factors issues in the cockpit and deliver that training program to every United pilot.

The United program was based on the pilot-human-factors research that had been done in the 1970s. It also drew from a commercially available business-management-training program. The course was, as I recall, five days long with most of the time spent in a classroom. A little simulator time was thrown in at the end of the course to bring all the things that were talked about in the classroom into focus in the cockpit. United's program was so successful in fact that it later sold generic versions of it to other airlines and aviation operations that were seeking ways to meet the FAA's new requirement for Cockpit Resource Management (CRM) training at the airlines. My role was to consult to the United development team on the best way to gain acceptance for this new approach to airline airmanship among United's pilots. The team knew that the radical changes that would be ushered in by implementing C/L/R were going to be hard to swallow for many of United's more-seasoned aviators.

I attended one of the first C/L/R courses held at the United Airlines Training Center in Denver, Colorado. There were forty-eight United pilots in the class—evenly divided among captains, first officers and second officers. About half way through the course, we were organized into sixteen three-pilot crews so that we could practice in a more-realistic setting the principles of C/L/R that we were being taught in the classroom.

Every evening, we would gather in the hotel bar and share our opinions on this whole new approach to airmanship. As one might expect, negativity to the program tended to come from the more-senior captains among us, with the more-positive end of the spectrum being represented by the junior first-and-second officers. During the hotel-bar debriefing on the last night we would be in Denver for initial C/L/R training, it became apparent that United was going to have some difficulty making a smooth culture change to what I now call Airmanship 2.0.

I reported my opinion to the development team and made a few suggestions on how to gain acceptance of the program by the resistors. I then returned my focus to the Space Shuttle opportunity and only occasionally met with the development team to offer my suggestions. It took close to a decade to complete the culture change. Many of the most intransigent nonbelievers remained so until their retirements.

However, I was a believer in C/L/R from the beginning. The initial training course and the follow-on recurrent training and education put me on a path to learning a lot more about the human-factors aspects of airmanship. I'm still on that path today.

The experience also convinced me that if a large aviation organization wanted to make a significant change in its pilot culture, the intense approach was much more effective than a more leisurely one. The culture changes I had designed and implemented (Operation USA and Operation Blue Skies) got significant results in approximately ninety days whereas United's rather low-key approach to implementing a pilot-culture change took almost a decade.

The complete story of this little-known episode in United Airline's history deserves more attention that we can give it here. It certainly provides insight into the often-traumatic changes the airline-pilot culture went through as the industry transitioned to a deregulated environment. I hope to be able to take the time to fully reveal that period in history from my perspective in a future book.

By the end of 1980, Percy Wood had become convinced that I had read the Space Shuttle opportunity right and that it was in United Airline's best interest to seriously pursue it. Dick Tabery, United's Senior Vice President Maintenance and Engineering had been attending several of the meetings I set up for Percy with NASA and various aerospace companies like Lockheed, Rockwell and Boeing. Dick was also of the opinion that United had a good chance of securing the opportunity and of successfully operating the Space Shuttle if we won it. I also arranged for Percy and the United team to be briefed by several high-ranking retired NASA managers and space-industry experts who could shed more light on the opportunity and how to go about capturing it. As we entered 1981, the momentum for the deal was increasing nicely and the circle of believers within United Airlines was expanding rapidly.

Unfortunately, there were a couple of holdouts—Dick Ferris (United's CEO) and John Ferg (the UAL-MEC Master Chairman). I had spent a lot of time with Dick Ferris while working on the Operation Blue Skies pilot-culture-change project. He generally encouraged me to keep working on the Space Shuttle deal, but I always felt that he had another agenda on that issue and that he was just charming me because he needed my help on the culture-change project. Ditto John Ferg. I could tell that there was still a strong undercurrent of skepticism running beneath the relationship that I had with both of them.

I probably had this feeling because both Dick and John had initially rejected my recommendation to look into the Space Shuttle opportunity. It wasn't until they decided that they needed to recruit me for Operation USA (the crew-complement campaign) that they changed their tune. And they continued to sing the new tune throughout the Operation Blue Skies campaign.

But the good news at this point was that Percy was enthusiastic about our chances of taking over commercial Space Shuttle operations from NASA, and the NASA folks were definitely courting United. Now, if NASA would just get the Space Shuttle Columbia off the ground for its first flight and complete its planned one-year orbital-test-flight program, maybe I could realize my dream of commanding regular Space Shuttle flights to low-earth orbit and back.

The first test launch of the Shuttle was set for early 1981. The first target launch date was in March, but a series of issues, including the deaths of two Rockwell technicians during a launch rehearsal, pushed that target to April 5th. Then the launch slipped to April 7th and then to April 10th. I had made arrangements with NASA for VIP passes for the first launch for Percy Wood, John Ferg and Dick Tabery. Dick Ferris declined my offer to join us. My Johnson Space Center media badge granted me access to both the media and the VIP areas at the Cape.

As the launch date slipped, I was kept busy coordinating everyone's schedules. In addition to getting the United team to the launch, I had made arrangements for a dinner party in Coco Beach the night before the launch. I had invited several current and retired NASA folks along with three or four aerospace-industry executives who I wanted to introduce to the United team.

In addition to maintaining my network of contacts at the Johnson Space Center (JSC), I had for the past couple of years been expanding my network at the Kennedy Space Center (KSC). I had been visiting the Cape fairly regularly to interview key NASA and aerospace-industry players for the book I was writing. My JSC media badge gave me, once approved by the KSC folks, access to KSC. However, unlike JSC, I normally had to be escorted by a NASA PAO (Public Affairs Office) officer. This apparent restriction actually worked in my favor since I got to know several of the PAO people well. They, just like the PAO staff at JSC, were always very helpful and eager to educate me about the spaceflight business. And they always knew who to take me to for the answers to my questions.

I guess the first Space Shuttle launch and an interesting dinner party was a pretty strong draw. Everyone on my list of invitees rolled with the delays like seasoned air travelers. So, on the evening of April 9th, 1981, Percy Wood, John Ferg, Dick Tabery and I attended the dinner party. Everyone, as I recall, had a great time. In addition, several good connections were made and a considerable amount of aerospace-industry intelligence was exchanged.

Percy, John, Dick and I left our hotel well before dawn the next morning and drove out to the NASA VIP viewing area that overlooked Pad 39A at the Kennedy Space Center. We met up with Sue Butler in the VIP lounge. Everyone had met Sue at the dinner party the night before. Sue had actually set up the dinner party for me. I can't recall if at the time she worked for NASA as a PAO officer, or she was still a space-beat reporter for a local paper. I can recall that at one point, Sue was also a public-relations executive for an aerospace company.

When Sue was a young girl, she escaped from Nazi Austria to England. She went to college there and ended up in the World War II OSS. Based on her wartime service, she was allowed to immigrate to the U.S. after the war. When I met Sue at some social function at the Cape in late 1979, she was intrigued about my plans for commercial spaceflight. It was my great good fortune that Sue took me under her wing and facilitated my establishment in the KSC community, just as Milt Reim had at JSC.

Sue had covered the Mercury, Gemini, Apollo and SkyLab programs for various newspapers. By the time I met her at the dawn of the Space Shuttle program, she was very well known in the space business as a knowledgeable, savvy and very-well-connected spaceflight expert. Sue was always ready to help me make a connection or find the right person to talk with. She introduced me to NASA veterans like Sam Beddingfield who had been with NASA since 1959 and was then the deputy director of Space Shuttle operations at KSC. Sam seemed to know everybody at the Cape. Sam and I went on to be good friends. With Sue's and Sam's help, by the time of the first Space Shuttle Launch, I was on a first-name basis with the KSC Director and his staff and the heads of most of the departments at KSC.

Sue made sure we all had coffee and then guided us to the VIP viewing area. As we settled into our seats in the VIP bleachers, we could see the **Space Shuttle Columbia** sitting on Pad 39A. It was bathed in bright-white floodlights. It looked like it was ready to go. The countdown was going smoothly and everything was on schedule for the planned 0650 EST launch. Unfortunately, at T-20 minutes a problem in Columbia's computer reared its ugly head. After several holds, the launch was finally scrubbed and rescheduled for April 12th.

Needless to say, everyone was disappointed. Percy, John and Dick had to leave the Cape to attend to other matters. However, they all planned to be back for the next launch attempt. I was able to stay over to get an inside look at how NASA was handling the relaunch process.

As it turned out, we had a little different team present at the KSC VIP area in the predawn hours of April 12th. Dick Tabery was unable to get back down to see the launch due to some pressing problems back at the United maintenance base in San Francisco. Percy and John had made it back and we were able to get VIP passes for Pat Austin (Vice Chairman of the UAL-MEC) and Chuck Pierce (Secretary-Treasurer of the UAL-MEC). Sue Butler and Sam Beddingfield were our guides.

The launch countdown went smoothly and at 0700 EST the Space Shuttle Columbia lifted off on schedule for the first of its planned four-flight orbital test program. The test-program plan had a one-year timeframe. When it was successfully completed, NASA planned to move quickly to turn over commercial Space Shuttle operations to the private sector. To me, that meant United Airlines.

After the launch, we hurried to Orlando to catch United flights. The enroute talk was upbeat all around. It looked like I was gaining traction with the ALPA hierarchy with the idea that United pilots could be flying the Space Shuttle in the not-too-distant future. I was certainly convinced that the opportunity we were pursuing could lead to that happy result.

Shortly after the first Shuttle launch, Capt. J.D. Smith, United's Senior Vice President for Safety and Industry Affairs, asked me to take a look at how United communicated safety information to its pilots. My report to J.D., titled "Safety Is The Ticket", outlined a comprehensive overhaul of United's internal communications to its pilots. J.D. adopted most of my suggestions.

By late 1981, the ALPA Professional Outlook Committee that I chaired had grown to over fifty members. Most of the members were highly motivated United pilots who had caught the Space Shuttle bug from me. We were all serious about preparing to fly the Space Shuttle professionally.

John Ferg pointed out to me that the committee was the largest ALPA committee he had ever seen and that I should rein it in to more reasonable proportions. After discussing options with John and conferring with Percy Wood, we decided that the best option would be to leave a drastically smaller committee in place and set up a separate (from ALPA and United Airlines) organization to prepare the United Pilots for commercial Space Shuttle operations.

Therefore, a few like-minded United pilots and I formed a non-profit professional organization that we named the **American Society of Aerospace Pilots** (ASAP). Dave Stoddard, Keeton Barnes, Art Ziccardi and Pat Pallazzolo were among the other key players who were founding members of ASAP. I was installed as the organization's Founding National Chairman. The organization's goal was to prepare its members for commercial-spaceflight operations and to ensure, through politics and public relations, that United line pilots would be flying the Space Shuttle when United took it over.

Our main motivation for doing this was that we knew that if we didn't do those two things, United would probably hire former NASA astronauts to do the job at a fraction of what we could earn under ALPA's collective-bargaining agreement with United. Traditionally, airline pilots have been compensated based on the weight and speed of the aircraft they fly. A **Boeing 747** weighs in at a little less than one million pounds at takeoff and cruises at about 560 miles per hour. The Space Shuttle weighed just shy of 4,500,000 pounds at takeoff and cruised at over 17,500 miles per hour.

As I recall, at that time a United 747 captain was making around $250,000 per year (over $700,000 in 2013 dollars). I figured that if we applied the weight/speed formula used to calculate a current 747 captain's pay rate, we could negotiate a Space Shuttle captain pay rate that would exceed $1,000,000 dollars a year. I could think of many reasons why a Space Shuttle captain should be paid four times what a 747 captain was being paid besides our traditional gross-weight-speed compensation formula. However the main reason I thought we could command such lofty paychecks was that the pilots we were preparing for Space Shuttle operations would have a corner on the commercial Space Shuttle pilot market. And the projected revenue that United could expect from each flight was around $150-million.

Of course, I saw myself as one of those captains. I believe it was at this point that I changed my airmanship goal from NASA Astronaut to United Airlines Space Shuttle Captain. However, as a backup I continued to politic for a NASA-astronaut slot with my NASA friends.

In 1982 and 1983, I continued to work as Percy Wood's special assistant for Space Shuttle acquisition. Upon NASA's recommendation, United had decided to team with Lockheed to bid on the first of five Space Shuttle operations contracts. We had private assurances from NASA leaders that the United-Lockheed team had every chance of winning it.

During this time, I also continued to serve as the ALPA UAL-MEC Professional Outlook Committee chairman and as the National Chairman of the American Society of Aerospace Pilots (ASAP). I was as well still serving on the publicity committee of the Houston chapter of the American Institute of Aeronautics and Astronautics and as president of the Houston chapter of the Aviation/Space Writers Association. I spent a lot of time during this period traveling around the country to speak at aerospace-industry conferences and make TV and radio appearances.

By 1984, ASAP had over a thousand dues-paying members (about 800 United pilots and 200 other ALPA-member pilots) who were enrolled in ASAP's Basic Spaceflight Ground School. Each ASAP member was signed up to pay $2,000 for the ASAP Basic Spaceflight Course. At this time, ASAP was attracting national attention through the numerous TV and radio appearances I made wherein I championed the concept of commercial spaceflight. Things were looking good on all fronts and my level of confidence that United would win the Space Shuttle operations contracts was following a trajectory that I believed would put my in the left seat of a Space Shuttle in the not-to-distant future.

I enjoyed working on all of the various aspects of United's campaign to win the NASA Space Shuttle contract. However, I liked most of all leading the team that designed and implemented the ASAP Basic Spaceflight Ground School. And, I especially enjoyed the challenge of creating and then instilling a new aerospace-pilot culture in ASAP and its members. As part of this effort, ASAP set up and ran special, NASA-supported tours for United pilots, flight attendants, mechanics, managers and other employees and their families to all of the Space Shuttle launches and landings from the second Shuttle flight through late 1985. Altogether, a few thousand United folks joined us on those and other outings to Space Shuttle related happenings.

By early 1984, United was positioned to win the first of five cost-plus contracts from NASA that would lead to the operation of all commercial Space Shuttle flights. Unfortunately, United was not awarded the contract. The story of this episode in the history of United Airlines deserves more space than I can devote to it in this book. I hope to tell it in detail in a future book.

It didn't take me long after United lost the bid to run commercial Space Shuttle operations to also realize that my chances of securing a NASA-astronaut job were all but gone along with my hopes of being the first airline-pilot astronaut to fly the Space Shuttle. Although I still had access to key

Chapter 1: My Pursuit of More Enjoyable Personal Flying

people in NASA and the aerospace industry, my reputation had lost a lot of luster because I was a highly visible part of the United team that had gone down a long road with NASA only to find that it led to a dead end. And I had made many statements during the course of traveling down that road that were not well liked by the powers in the NASA astronaut office.

I soldiered on with the ASAP crew for a year or so after that, hoping to find an opening for airline pilots to transition to commercial Space Shuttle operations. However, with the January 1986 Challenger disaster, I lost all hope of realizing my dreams of taking my airmanship to literally the next level. The good news was that I believe I proved that airline pilots could make that transition.

In the early fall of 1984, I went through one of United's first **Boeing 737-300** qualification courses as a first officer. This was my introduction to glass cockpits. Up until this point, I was strictly a "steam gauge" pilot. I've always been an early adopter of new technology and I really enjoyed learning this new level of airmanship. I became a believer in glass cockpits in the early 1980s and I'm still one today. My status at United soon reverted to that of a regular line pilot flying as a 737 first officer. To make matters worse, it looked like it was going to take another five-to-ten years for my seniority to move up to the point where I could capture a captain bid.

My long-range plan at this point was to build up my public-relations/corporate-communications consulting company (AeroComm) to the point it could support my family. While I was doing that, I planned to grab the first captain bid my seniority would buy me, check out as a captain and then retire from United. I would have left the company in 1985, but I had signed on with United to fly as a captain and I wanted to achieve that airmanship goal before leaving the company.

In late 1984, I was invited by the United Airlines ALPA leadership to design and implement a large-scale pilot-culture-change project for the United-pilot group. The goal of the culture-change project was to create a then-non-existent solidarity among the United pilots so that they could prosecute a successful strike effort if called upon to do so. I at first declined this invitation because at this point I did not intend to be with United much longer and I was busy building my business.

ALPA wanted to get ready for a strike because the intelligence it had was pointing to the very real possibility that Dick Ferris was planning on forcing the United pilots out on strike and then breaking the pilot union at United, just as Frank Lorenzo had recently done at Continental Airlines. It wasn't very long after the ALPA leadership approached me that I learned from my own trusted

61

sources that Ferris had indeed received his marching orders from United's board to destroy ALPA at United.

After a second approach by the ALPA leaders, I decided to undertake the assignment. I accepted it because I believed that if ALPA were broken at United, the other airline-pilot groups in the organization (Delta, Eastern, etc.) would also be broken. This would lead to a general and rapid deterioration of the profession. I didn't want to see this happen because I felt that I owed a lot to the airline-piloting profession and I was at that time beginning to plan a new business that was focused on preparing young people for a career as an airline pilot.

However, there were some caveats to my agreement to take on the project. The first stipulation was that although I would volunteer my time to the effort, the subcontractors who worked for AeroComm would charge their normal fees and I would receive my normal agency fee from them. Second, I wanted the authority from ALPA to set up and run an ALPA committee that would design a buy-out of the airline by the United pilots. Since the deregulation of the airline industry, I had been advocating that if airline pilots bought their airlines (they had the financial power to do so), we could avoid most of the unpleasant effects on our profession that were being caused by deregulation.

By this point, only a few short years following deregulation of the airline industry, the workload for United's pilots had approximately doubled and our compensation level had been reduced by around thirty percent. Given the dynamics of the "new" airline industry and the thinking of those at the top of the ALPA hierarchy, I concluded that the airline piloting profession would continue to become less desirable unless we took control of our destiny. The ALPA leadership agreed to both of my requirements.

The AeroComm team did a very effective job of helping ALPA to solidify the United pilot group. This was possible because I expanded on what I had learned in my earlier culture change projects—Operation USA and Operation Blue Skies. We of course couldn't use the video-viewing systems that had been installed in the pilot lounges around United's system although they were still in place and being used by United to communicate to the pilot group. I therefore created what came to be known as the "United Family Awareness Program". It was based on the concepts that all United families must have access to up-to-date and accurate information from their leadership, all families must be able to let their leadership know how they feel and have confidence they are being heard and all families must be able to get support from each other during the difficult times that would result from a strike.

We recruited and trained one pilot-spouse team for every ten United pilots. The pilot-spouse team then convened a meeting in their home every time we produced a new video that addressed the important issues related to the possible strike. The other nine pilots in their group, along with their spouses and/or significant others, attended the meeting to view the videos, discuss the issues and then provide structured feedback to the ALPA leadership. We also greatly expanded our use of video teleconferencing. The productions were bigger and better done, there were more of them and they were attended may many thousands of United pilots, their families and other United employees.

When everyone's worst fears were realized and the United pilots went out on strike on May 17, 1985, ninety-six percent of the United Pilots chose not to cross the picket lines. This percentage increased as the strike wore on. ALPA was so pleased with the results that they retained AeroComm in late 1985 to do the same thing for the Eastern pilots when Frank Lorenzo took over the airline.

I chaired the ALPA-authorized United Pilots Buy-Out Committee before, during and after the 1985 strike. We produced a buy-out plan that did not include an Employee Stock Ownership Plan (ESOP). In fact, the committee had determined that an ESOP would be the worst way to acquire the company because of the many pitfalls inherent in an ESOP. We, of course, reported our findings to the ALPA leadership, but for political reasons ALPA decided to use an ESOP as the pilots' vehicle for acquiring United. ALPA eventually implemented the ESOP strategy to acquire United in 1994.

I did not agree with the ESOP approach and resigned from the committee because I believed it was what pilots refer to as a "sucker hole". That's a hole in the clouds below you that you decide to use to try to get below the clouds. Many pilots have come to a bad end trying to get to visual conditions below the clouds through a sucker hole. I was very unhappy to see that this ESOP was indeed a sucker hole that manifested itself when the United pilots lost all their equity in United, and a large portion of their retirement savings, when United went through bankruptcy from 2002 to 2006.

In late 1986, I developed a long-range, multi-tiered business plan. The first part of the plan was embodied in the Aerospace Flight Training Academy (AFTA). AFTA was a not-for-profit company that focused on providing career-guidance information for aspiring airline pilots. The backbone of this program was an audiogram (audio-cassette tapes and a workbook) that I wrote and produced titled "Jet Airline Pilot: What It Means To Be One / How To Become One". AFTA sold the audiogram directly to young people who wanted to pursue a career as an airline pilot. We also provided personal consulting to these young folks and their parents.

Because I needed a more-capable airplane by this point for my business- and-personal-mobility needs, I sold my Cardinal and Yankee and purchased a **Cessna 337 Skymaster**. It is a four-seat, pressurized, twin-engine airplane that allowed me to stretch my personal wings on missions that didn't require airline travel. I also bought another Stearman for my fun flying.

By late 1987, I had confirmed that there was a considerable market for not only aviation-career guidance information, but also for quality flight training that focused on preparing young pilots for the airlines. I therefore set to work on the second phase of the business plan. This phase included the establishment of a professional flight-training academy that I named "Aerospace Regional Training Academy (ARTA). ARTA was designed to be the prototype for a national network of flight-training academies that would prepare aspiring airline pilots to be the most-competitive candidates for the thousands of airline pilot jobs that were opening up due to airline expansion and pilot retirements.

I designed the structure of the business and developed all of the ground-and-flight-training courses in the ARTA career-track program. The career-track program was designed to qualify a candidate who was a zero-time pilot and high-school graduate to fly for the airlines, and to provide them with an Embry-Riddle Aeronautical University Bachelor of Science in Aeronautical Studies degree. Our program allowed our students to complete our program, including the BS degree, within three years from that time they started the full-time program with us.

I launched the prototype ARTA in Grand Rapids, Michigan in early 1988. We presold the program and before we opened our doors we had about fifty full-time students. By the end of 1989, we had over seventy students in the program and things were working out pretty much as planned.

The training program included several flight-training innovations that I designed and implemented. One was to require the students to brief their flight instructors before every flight instead of the flight instructors briefing the students. I had learned long before this that the best way to learn something is to teach it. So, instead of students half listening to the instructor's briefing, the students had to concentrate on the material in the briefing because they had to teach it to the instructor. The instructor would then critique the student's briefing.

Another innovation that I invented and introduced was the video recording of all flight lessons. This was back in the days of VHS tape recorders and comparatively large video cameras. The system included two video cameras in the cockpit—one focused out the windscreen and one on the instrument panel. The video from these two cameras was fed to a "splitter box" that then fed two video tracks to a VHS-tape recorder. The recorder produced a video with the view out the windscreen on the top third of the image and the instrument panel on the bottom two thirds. Audio from the onboard intercom system (including

Chapter 1: My Pursuit of More Enjoyable Personal Flying

all radio communications) was synchronized with the video to produce a complete, seamless recording of every training flight.

This proved to be a very valuable training tool. The students and instructors used it in the debriefings for every flight and the students could take the tapes home and replay the training flight as many times as they wished. My chief flight instructor could also use the tapes for quality control and standardization. Modern versions of this system are readily available now, but as far as I know, this was the first use of such a system in civilian pilot training. Unfortunately, most flight schools today still do not use a video-recording system for flight training.

The Grand Rapids ARTA also sported a Frasca flight simulator that was fully integrated into the training program. At that time, very few civilian flight schools used simulators. Unfortunately that is still the case today.

I designed and installed a unique pilot culture in ARTA to provide the students with the airmanship-development attitudes they would need in their personal pursuit of airmanship excellence. It was a blend of military, airline and general-aviation pilot cultures with heavy emphasis on the military-pilot culture. For example, all students and instructors wore military-style flight suits and military-style flight helmets in the cockpit. The development of the culture was facilitated by frequent social gatherings and "enrichment" opportunities like group flights to points of aviation interest and participation in airshows.

In late 1988, I took a month off from my business endeavors and went to United's **DC-10** initial-training program in Denver as a first officer. By this time, I could see a captain bid on the horizon maybe only a year or so away. It was common wisdom in our United-pilot culture that an airman should fly a widebody jet as a first officer before he checked out on one as a captain. Since my captaincy was approaching, this would be my last chance to follow this wisdom.

I really enjoyed the DC-10. The flying was considerably different than what I had experienced on the 727 and 737. It included longer legs and a lot of over-water flying on the Hawaii runs. I always felt that I was flying a fully loaded freight train through the sky at 600 mph when I was at the controls of the '10.

By early 1989, the prototype ARTA in Grand Rapids was breaking even, the students and their parents were happy with the program, and the student cadre was growing by about a dozen new students per month. I determined that this phase of the business plan had verified my assumptions about the viability of a national network of ARTAs. So, I set about developing a detailed business plan for the next phase.

65

I completed the new business plan within a few months. It included expanding the Grand Rapids ARTA, purchasing an aviation-support facility (FBO) on the Grand Rapids airport, purchasing a flight-simulator manufacturer and then establishing ARTAs in five additional locations around the U.S. Once this phase was complete, the plan called for expanding the network to the top fifty cities in the U.S. (by population).

By the fall of 1989, I had purchase contracts on the Grand Rapids aviation-support facility, Link-Miles Simulators and SimuFlite. Link-Miles was a unit of the Singer-Link Company. Singer-Link was a conglomeration of the old Singer sewing machine company, the original Link simulator company and several other related and unrelated companies. When I arrived on the scene, the junk-bond artist who had engineered the acquisition of Singer Link a little more than a year earlier was selling off the various units of the Singer Link Company. Link-Miles Simulators and SimuFlite were that last two, as I recall, units on the block.

At the time, Link-Miles Simulators was producing airline and business-jet simulators. The military-simulator unit had already been sold to CAE of Canada. SimuFlite was the world's second-largest corporate-pilot-training vendor. It had about twenty-percent of the global corporate-pilot-training market, second only to FlightSafety International.

With the acquisition of SimuFlite, I would have a major facility in Dallas with airline and corporate-aircraft simulators. My business plan called for this central training center to be fed students by the network of ARTAs. General-aviation-aircraft-type simulators, produced by Link-Miles Simulators, would be located at the ARTAs. Link-Miles Simulators would manufacture all of the airline, business-jet and general-aviation-aircraft-type simulators needed for the ARTAs and SimuFlite.

Up to this point, I had self-financed all of my business ventures after my first start up in 1967. However, the hefty price tag on the business acquisitions in the new business plan necessitated that I bring some investors into the deal. I obtained the assistance of an investment banker and by late 1989 we had commitments for the financing we would need to close the deals for the aviation-support facility, Link-Miles Simulators and SimuFlite. An additional capital contribution from me would cover the operating capital we needed to implement the new business plan.

In September of 1989, I returned to the United Airlines Training Center in Denver for initial Boeing 737-300 captain training. I completed the training in October and shortly thereafter started flying the line as a newly minted captain almost exactly twenty-one years after my first flight as a United second officer back in 1968. I had finally reached my airmanship goal of airline captain.

So, in late 1989 I reached another major decision point in my life. I had achieved my stated airmanship goal of flying the line as a captain. I had a profitable business in the ARTA in Grand Rapids and I had a financed business plan ready to go that I was very anxious to implement. There was only one fly in the ointment that I was aware of. As it turned out, another very big fly was hiding in there too.

The known "fly" was the fact that the investors I had lined up told me that they would close on the financing for the deal only if I resigned from United and gave my full focus to the new venture. I fully understood their concerns and I had agreed to those terms. However, I was hoping to put off my resignation from United until the summer of 1990. I was really enjoying flying as an airline captain and wanted to get a little more of it before I hung up those particular spurs. Also, as a typically cautious pilot, I would have preferred to hang on to the United paycheck until I was sure the new venture was going to be successful.

1990s

As it turned out, the investors, the investment banker and some other factors brought me to the decision in early January of 1990 to tender my resignation to the company that I started with as a kid and where I had always hoped I could retire from with a full airline-pilot career. As it was, I had completed a little over twenty-two years of my projected thirty-nine year career with United Airlines—a run that I am now more than satisfied with. However, at the time this was one of the hardest decisions I had ever made. I loved actually flying airliners although by now I was almost completely dissatisfied with most of the other aspects of the job. On the other hand, I truly believed in the business plan I was about to launch and my ability to implement it. In the end, I resigned from United and faced the new decade with an abundance of optimism and energy.

Unfortunately, the other fly hiding in the ointment revealed himself shortly thereafter. When Saddam Hussein (the hiding fly) started rattling his saber over the Kuwaiti oil fields in early 1990, the investors I had lined up decided that they'd better wait to see how things turned out in the Middle East before closing our deal. As things dragged on, I was able to get extensions on the closings of the Grand Rapids aviation-support facility, Link-Miles Simulators and SimuFlite deals. However, when Saddam's army rolled into those Kuwaiti oil fields in August 1990, everyone, including my investors, thought that Armageddon was nigh. By the end of the month, the U.S. and its allies had launched Operation Desert Shield and what turned out to be the massive buildup to the First Gulf War. Shortly thereafter, the investors formally backed out of the deal pending the outcome of world events, and I had to let the purchase agreements for the Grand Rapids aviation-support facility, Link-Miles Simulators and SimuFlite expire.

I hunkered down to watch the military buildup and maneuvering on CNN while my team at the ARTA in Grand Rapids kept the business running. For a while, things were fairly stable, but as the months wore on, our new enrollments started to fall off alarmingly. At this time, a growing number of our students were being pulled from the program by their parents because of the uncertain future facing their families. If you were around at that time, I'm sure you recall what the general state of our society's collective mind was—*the world is coming to an end!*

By the time the First Gulf War ended in early 1991, I had been forced to shutter the Grand Rapids ARTA after pouring a sizeable portion of my operating capital into keeping it running. I, of course, didn't want and had not anticipated this outcome. On the plus side of the ledger, I had learned a lot about the flight training business and how to put together business-acquisition deals. I had also proven several of my pet hypotheses about how to recruit and train pilots and instill in them the pursuit of airmanship excellence. And I still had what I believed to be a viable business plan. On the minus side, I was no longer receiving my generous United-captain's paycheck on the first and fifteenth of every month and I had expended a considerable portion of my financial reserves.

Unfortunately, the pressures that built up around this chain of events resulted in some major changes in my personal life that took me a couple of years to work through. The process and its consequences diverted my attention from trying to relaunch the Aerospace Flight Training Academy business plan until 1994. From early 1991 to mid-1994, I licked my wounds, did a little corporate flying, took on a couple of consulting assignments and plotted my comeback. I must admit that this was a very trying time for me. One I would not care to go through again and I wouldn't wish the experience on anyone. However, these tough trials generally do make us stronger and better equipped to face the future—if we survive them. In this case, I survived, but as it turned out just barely.

In 1991, I took on a consulting job for the owner of a fixed-base operation (FBO) at the Frankfort Airport located in Frankfort, Illinois. FBOs provide services (like fuel, storage, maintenance, flight training and charter) at most of the over 5,000 airports in the U.S. The owner had just purchased the FBO and since he was only a private pilot with no aviation-business experience, he asked me to produce a full business plan for him.

I researched and wrote the plan. It contained marketing plans for each of the five departments in the FBO. The marketing plan for his flight school contained the establishment of an aviation club in each of the high schools within a thirty-minute drive time of Frankfort Airport. I had started thinking about this approach to flight-training marketing when I was running the Grand Rapids ARTA.

After reviewing the business plan, the FBO owner retained me to help him implement it. As part of the implementation effort, I established aviation clubs in twelve high schools in the Frankfort Airport market area. Unfortunately, before we could fully implement the business plan, the owner decided, wisely I think, to sell the FBO since he had a full-time job as a college administrator and he really didn't have the time and money to do justice to the FBO business.

I also took on a consulting assignment for an FBO located on the Rockford, Illinois airport. In this case, the FBO owner had been in the aviation business for many years, but his FBO survived on revenues from only a charter department that specialized in cargo. He was also authorized to sell fuel, do maintenance work and have a flight school. Up to this point, he had not tried to capitalize on these additional authorizations, and he wanted me to create a business plan to do that. I produced a plan and handed it over to him to implement.

In 1992, I was doing some flying for **AAR Corp**. They had a **Challenger-600** and a **Cessna 421**. I flew copilot on the Challenger for a few trips (one to Europe), but I mainly flew as a safety pilot on the 421. The 421 was Dick Tabery's personal plane. Dick had left his job as United's Senior Vice President for Maintenance and Engineering around the same time I left United. We had stayed in touch after the Space Shuttle deal fell through. He was now AAR's co-chairman. Dick asked me to fly with their Challenger captain for a while to evaluate his performance. And he invited me to fly with him in the Cessna 421 as his safety pilot.

Dick was a former U.S. Army aviator of Korean-War vintage and an accomplished aviator although he had not been a line pilot at United. Because Dick had been a military pilot and he knew the intimate workings of United's flight operation, he was vividly aware of the potential dangers an aging aviator can face and the ways to mitigate the risks associated with those dangers. Since he was in his early sixties and he didn't fly enough on a regular basis to be proficient in a high-performance-twin, single-pilot operation, Dick always flew with an experienced safety pilot.

Since I wanted to explore the possibility of teaming up with AAR to relaunch the Aerospace Flight Training Academy, flying with Dick gave us the opportunity to talk about it at length. And we enjoyed flying together. Things were going along well until a dark, cold night in December 1992. That's when I experienced my first and to-date only airplane crash.

Early in the morning on the day of the accident, I flew out of Chicago Executive Airport (then named Palwaukee Municipal Airport) as a safety pilot for a business associate in his **Cessna 402**. We flew down to Springfield, Illinois to meet with the CEO of Midway Connection. The regional airline was in bankruptcy and we were interested in purchasing some of its assets for a start-up regional airline that was part of the new business plan I was working on. The new plan essentially merged the flight-training academy concept with a regional airline that would fly from reliever airports in major metropolitan areas to reliever airports in other major cities. Thus bypassing the increasingly overburdened airline airports. The graduates of our flight-training academy would fill the co-pilot seats on these flights.

The night before the flight, I had been out late visiting with friends and I had quite a bit to drink. Because of the "oh-dark-hundred" launch that day, I got very little sleep the night before. I knew that I shouldn't fly at all that day because I was fatigued and hung over, but the fact that I was essentially riding along in the copilot's seat lulled me into a false sense of security. After all, since I wasn't going to be the Pilot In Command, there was no legal requirement for me to be fit for duty.

We had two passengers back in the cabin of the 402. Robert "Jorgey" Jorgensen and Colin Murray. Jorgy started his aviation career as a B-25 test pilot in WWII. After the war, he spent twenty years with Northwest Airlines and retired as the Director of Flight Engineers. He had also served as the Vice President of Engineering for Hughes Airwest, Executive Vice President of Philippine Airlines, COO of Horizon Airlines, COO of Midway Aircraft and Vice President of System Safety for Midway Airlines.

Jorgey was helping me put the planned regional airline together and he knew the CEO of Midway Connection, Richard Pfennig, very well. Colin was recently retired from United Airlines after serving as the company's Regional Vice President for the North Pacific. He was advising me on marketing and financing for the proposed airline.

The flight to Abraham Lincoln Capital Airport in Springfield was uneventful. We spent the day meeting with the Midway Connection folks. Late in the afternoon when we were about to return to Chicago, I got a call from Dick Tabery. He was in Oklahoma City at one of AAR's fixed base operations (FBOs). The 421 had just come out of its annual inspection and Dick asked me if I would come out to Oklahoma City and fly it back to Palwaukee Airport with him.

I at first demurred, citing my long day and the difficulty I would have in getting out to Oklahoma City at a reasonable time. Dick pressed me to agree to the flight. So, I talked with the owner of the 402 about returning to Chicago in the 402 as soon as we wrapped up our Midway Connection meeting. As it turned out, the owner of the 402 had planned to fly it out to Tulsa after dropping us off at Palwaukee Airport. He had business in Tulsa the next morning. He graciously offered to drop me off in Oklahoma City on his way to Tulsa. Jorgy and Colin agreed to come along because they both knew Dick well and they wanted to help him out.

I called Dick back and told him I planned to arrive in Oklahoma City around 9 PM local time. We finished our meeting in Springfield and flew the 402 out to Oklahoma City. By the time we arrived, Dick had wrapped up his meetings and he had the 421 ready to fly back to Chicago. We launched around 10:30 PM.

Every accident has a chain of events that lead up to it. There are usually three-to-five links in what we call the "error chain". By the time we took off from Oklahoma City, there was already more than one link in the error chain that led to our subsequent accident.

It was a beautiful, cold, starry night and we enjoyed a lively four-way conversation as we sailed along at our cruising altitude of 22,000-feet munching on the catered sandwiches Dick had boarded for us. As we approached the Chicago area, Chicago Center cleared us to descend to a lower altitude. Shortly after we started the descent, we passed quickly through a very thin layer of clouds. Both Dick's and my windscreens immediately iced over. It was impossible for us to see anything out in front of the airplane.

Dick turned the windshield-anti-ice system on. Unfortunately, on this 421 the only thing deiced by the system was a five-inch-high strip that ran along the bottom of Dick's windshield. My windshield remained completely opaque. Since the temperature on the ground was well below freezing, we expected the ice to remain on the windshields until we got the airplane back in AAR's heated hangar at Palwaukee. Another link had just been added to the error chain.

Now out in clear air, we could see the entire Chicago area spread out below us through the cockpit's side windows. Chicago approach rushed us to cancel our instrument clearance and proceed to Palwaukee on our own. I told Dick that I had Palwaukee in sight out of my right-side window and he decided to cancel our instrument flight plan and proceed to the airport in visual conditions. The Palwaukee tower was closed because it was now after midnight. Dick's decision to cancel our instrument flight plan and the fact that the tower was closed added two more links to the error chain.

We had briefed a landing on Palwaukee's Runway 16 since the wind was out of the southeast. That runway was equipped with an Instrument Landing System (ILS) and Dick and I both tuned it in on our navigation radios and set our flight instruments for the approach. When Dick turned in towards the runway, I could no longer see it since it was now out in front of us. At this point, my state of fatigue was approaching a critical level and I more or less relaxed while Dick made the landing approach. I thought at this point that we had it made. It was a clear night with a very light wind and Dick was landing an airplane that he was very familiar with on a runway he had landed on hundreds of times.

Things progressed smoothly until we were over the end of the runway. My peripheral vision told me that Dick was leveling off about fifty-feet in the air when he should have continued his descent to the runway until we were only a couple of feet above it. Shortly thereafter, he reduced the power and the stall-warning horn started to blare. If we entered a full aerodynamic stall at that height above the ground, there was little chance that we would avoid a serious crash.

I yelled "power, power, power" and got on the controls with Dick. We managed to keep the wings level and the airplane from entering a full stall. However, we hit the runway very hard and bounced back up into the air. We then managed to reduce the power and we touched down much more smoothly the second time. Unfortunately, the left-main landing gear had been broken on our first touchdown and it failed on the second.

We went off the left side of the runway at about eighty knots. As soon as we hit the snow, the nosewheel and right-main gear collapsed. Fortunately, there was a couple of feet of snow on the ground and the area to the left of the runway was clear of any hazards that could bring us to an abrupt stop. In fact, the deceleration was very smooth and we were able to run through our emergency-shutdown checklist as we slid to a stop.

After stopping, my airline-pilot training kicked in and I started yelling, "unfasten your seatbelts and get out". Since all four of us aboard were veteran airline people, there was no delay in exiting the airplane. Luckily, no one was injured and we moved well away from the airplane since there was smoke billowing up from under the left engine. By the time we got a few hundred feet away on the upwind side of the airplane, the smoke had disappeared.

We called the local police and fire department and stood out in the cold night air trying to calm down and assess the situation. The airplane was sitting on its belly in the middle of Runway 16 pointed in a northwesterly direction. This was confusing since we thought we went off the left side of Runway 16 heading in a southeasterly direction, sliding for several hundred feet before coming to rest. We backtracked our skid marks in the snow and determined that we had actually touched down on Runway 12, a shorter and narrower runway. We had indeed gone off left side of the runway and slid through the snow until we came to Runway 16 whereupon the airplane ground to a stop and rotated about ninety degrees to the left to end up pointed northwest.

After securing the wreck and notifying the FAA about the accident so they could close the airport to other traffic, we remained a safe distance from the airplane and debated whether or not we should hike the half mile or so to the AAR hangar or wait for the fire trucks to arrive. While we were considering these options, we noticed an airplane lined up for a landing on Runway 16 with its landing lights on. It appeared to be a corporate jet. Since the tower was closed, there was no way to communicate in a timely manner with the unsuspecting pilot of this airplane that would soon be barreling down Runway 16 unaware that the 421 was sitting on the runway. We didn't even have a flashlight. All we could do was watch helplessly and hope that the pilot would pick up the 421 in his landing lights with enough time to stop his jet before it ploughed into the 421. We moved farther away from the runway and hoped for the best.

As the airplane touched down on the northwest third of Runway 16, we identified the airplane as a small-cabin **Cessna Citation**. The airplane decelerated smoothly and as soon as the pilot saw the 421 in his landing lights, he applied full brakes and came to a stop only twenty-feet or so away from the 421. After a short pause, the pilot turned the Citation around on the runway and taxied off to his hangar on the northeast side of the airport.

The fire trucks and the police arrived shortly thereafter. They secured the airplane and placed flashing hazard lights around it before giving us a ride to the AAR hangar. When we got to the hangar, we called the FAA Flight Service Station in Kankakee, Illinois and made sure they issued a notice to airmen (NOTAM) that alerted other pilots to the fact that there was a wrecked airplane sitting on Runway 16.

After that, Dick put on a pot of coffee and the four of us debriefed the accident sequence. It was obvious that we had allowed several links to creep into the error chain that led to the accident. If we had removed even one of those links, the accident probably would not have happened.

Both Dick and I should not have even flown the flight because we were both fatigued from a long day of doing business. When the windshields iced over, we should have discussed our options and decided to land at an airport with an open control tower, crash-rescue equipment and a longer and wider runway. We should have also discussed the sight picture Dick would have when he raised the long nose of the 421 in the landing flare with only a five-

73

inch-high strip of clear windshield to look through. If we had made a decision to land at O'Hare Airport, DuPage Airport or even Milwaukee General Mitchell Airport, we could have avoided the problems we ran into trying to land at Palwaukee Airport. We also could have done a much better job of backing up our approach to Runway 16 with crosschecks of the ILS.

As it turned out, we were lucky. Although the airplane was totaled, no one got hurt. Some of that was due to good airmanship during the crash sequence, but there was no denying the breakdown in our airmanship before we reached the final link in the error chain that caused the hard landing on Runway 12. Needless to say, I went over the flight in my mind many, many times in the weeks that followed. Jorgey and Colin contributed their recollections of the crash sequence and their wisdom on how we handled the situation. And of course, Dick and I went over the details of the flight many times.

This was the first time that I'd been directly involved in an airplane accident. At the time of the accident, I was forty-five, I'd been flying for over thirty years and I had accumulated over 17,000 hours of flight time. I'd been trained by the best airmanship-trainers on the planet and I was very well versed in the techniques and methods that an excellent airman uses to identify and mitigate the risks we faced that night. Dick had over 5,000 hours of flight time and he had received his original airmanship training in the Army. Nevertheless, we had allowed an error chain to build up to the point where it resulted in a potentially fatal accident.

I learned some very important lessons from the experience. The biggest shock was the realization that it really could happen to me. This was, after all, the first aviation accident I had participated in. More importantly, I learned that I was not invincible. Before long, I had revamped my personal minimums based on a reassessment of the risk-reward equation I had been using up to the time of the accident. The experience also renewed my interest in the human-factors challenges pilots face. And it gave me some useful insights into how to support pilots who are using a personal aircraft for their business-and-personal-mobility needs.

The airline deal Jorgey, Colin and I had been working on was based on a concept I'd been thinking about since the early 1980s. It became obvious to me shortly after airline deregulation that the traditional network-airline system in the U.S. would become terribly inefficient and more like a bus service than the premium-airline system we had before deregulation. And in my vision of the future for airline travel, I couldn't see things trending in any direction but down in terms of the airline-travel-value proposition. After all, the biggest constraint to any airline system is the "land-side" facilities—runways, terminal, parking, access, etc. All of these facilities are planned, contracted-for and paid for by an intricate cabal of special, local, county, state and federal government agencies.

It has been the case for as long as I've been in aviation that the land-side capabilities of the U.S. airline system have always been way behind the demand placed on them. Since I had always known this to be so, and I couldn't see anything in the cards that was likely to improve the situation, I naturally formed the hypothesis that given the increasing demand for airline travel that deregulation was bringing about, the value proposition offered by the U.S. airline system would continue to depreciate in the view of a growing segment of the air-traveling public.

I came to that conclusion because I knew from my by then twenty-seven years in the air-transportation business that people bought primarily two things when they paid for their air travel: time and the experience. I believed then, and I believe now, that these two components of the air-travel-purchase decision are by far the most important for most people. No matter what type of air travel they're buying. I strongly believe that of these two components of the air-travel-purchase decision, time is of far greater value to most business-and-personal travelers in most cases. But airline portal-to-portal-travel times were at that time increasing (and they still are).

Here's a factoid that neatly illustrates the point. From the time that our ancestors first climbed on a horse right up to airline deregulation in 1979, portal-to-portal travel times had been *decreasing*. After 1980, portal-to-portal travel times started *increasing* for the first time in human history. In the early 1990s, the trend was still in that direction with nothing on the horizon to suggest that things might change anytime soon (and they haven't yet).

Most of us place a very high value on our time. And when we're squandering time on things like traffic delays (air and ground), airline-domino delays and going some place we don't want to go to get to where we do want to go, we all get frustrated and stressed out. That's because we know that this wasted time could be put to much better use with our families, our friends and our business associates. And that's the most valuable time of all.

The increased frustration and stress also corrode the air-travel experience. Of course, the devaluation of the "experience" is also the result of declining service levels and the lackluster attempts by most airlines to genuinely add value to the contemporary airline-travel experience. Being treated like cattle doesn't enhance the experience for most of us. And don't get me started on the TSA.

My vision of the future of legacy airline travel was juxtaposed with my experiences as a personal flyer, charter pilot and corporate pilot. Air travel outside the airline system has always saved time over any other travel mode and the experience of private-aircraft travel for most of us has been both very special and very satisfying. And for personal flyers, the rewards garnered from being able to fly yourself anywhere, anytime you want adds tremendous value to the experience.

By 1994, the plan based on the above thinking had evolved into what I called the "Cloud 9 Air Travelers Club". Members of the club could learn to fly in club aircraft and then fly club aircraft for their business-and-personal mobility needs. When they weren't flying the aircraft themselves (because they weren't qualified to fly the type of flight they wanted to make or they just didn't want to fly for various reasons), they could fly on flights arranged for them by the club that were flown by professional pilots.

The key business-plan element was a network of private-jet flights that fly on a schedule between reliever airports. For example, a private-jet flight that departs in the morning from Chicago DuPage Airport and flies non-stop to Westchester County Airport in the New York area. That same airplane then flies back to Chicago DuPage Airport in the evening. The seats are sold individually—just like buying an airline ticket. I believed this strategy would open the market for private-jet travel to a much larger number of people who can't justify or afford private-jet travel when they have to pay for the whole aircraft.

We planned to service this newly opened market with a regional airline that would initially fly out of a Chicago-area reliever airport to fifteen reliever airports in other metropolitan areas on an "out-in-the-morning, back-in-the-evening" schedule. Once the Chicago operation proved viable, we planned to expand to fifty additional cities. There would also be an on-demand charter option.

The plan called for a Cloud 9 Air Travelers Club aviation-support facility on a Chicago-area airport. This facility would include hangars for the private jets, maintenance bays, a flight school and a clubhouse for Cloud 9 members. It would also house a fleet of small, general-aviation aircraft for the flight school and the business-and-personal-mobility needs of the members.

Cloud 9 Air Travelers Club members would be able to avoid the pain of airline travel when flying to major cities in the U.S. at a price only slightly higher than airline first-class travel. Those members who wanted to fly a private aircraft themselves would learn to fly in the club's flight school. The flight school would emphasize airmanship rather than just learning to drive an airplane. Small, private aircraft would be available to graduates of the flight school through an aircraft-rental operation.

I rounded up some investors and a launch team, and stood Cloud 9 Air Travelers Club up at DuPage Airport in West Chicago, Illinois in the summer of 1995. We started by opening up the flight school and aircraft-rental operation first because they required far less capital than the private-jet airline. The plan was to get that part of the business going first and then bring in new capital to stand the airline up.

By late fall, the business was profitable and we were adding members at a geometric rate with the help of a very effective member-referral program. The former U.S. Air Force **T-38** instructor pilot I hired as chief instructor did a great job of helping me to introduce a professional airmanship culture into the club. The entire Cloud 9 team did a great job of modeling and upholding very high professional and airmanship standards. Everyone involved in the Cloud 9 operation—members, staff and potential members—actively participated in the airmanship culture we created. Most importantly, everyone had a good time.

In late 1995, I went back to the investment advisor who represented the largest investor group. We had a gentlemen's agreement that he would secure the needed expansion capital from his group of investors if we hit our start-up milestones. I told him that we had reached or surpassed all of the start-up goals in our plan, and that we were now ready for the next round of financing that would rapidly expand the club and lay the groundwork for the launch of the regional-airline operation.

After considerable discussion with the investment advisor, he made me an offer I couldn't refuse. He proposed to buy, on behalf of one of his clients, my equity in the company. At first, I was reluctant to accept the offer, but some things that were happening in my personal life biased me towards accepting it. And I had after all done what I enjoyed best—creating and launching a new company. I knew from prior experience that although I can (and have) run businesses on a day-to-day basis, I soon tire of the repetition involved and start to yearn for a new entrepreneurial challenge. In the end, I accepted his offer near the end of 1995 and went into what I hoped to be a semi-retirement phase. I planned to spend a lot of time lounging in the sun and writing a book that had been percolating in my mind since the early 1980s.

As a result of my research into spaceflight and what I learned during my stint in the Space Shuttle-acquisition deal, I had become aware of a strategy that I believed would, if implemented properly, provide the means to safely, routinely and affordably travel from the Earth's surface to low-earth orbit and back. My intent was to explain this strategy in the book I intended to write.

I rented a beach house on the southeastern shore of Lake Michigan. It had been built in the late 1800s among the tree-capped sand dunes located about two hundred feet from the lakeshore. It had a huge stone fireplace and I settled down in front of it to write the book. The winter of 1995 - 1996 was unusually cold and my location on the downwind side of Lake Michigan ensured an ample supply of snow. So, it wasn't too hard to convince myself to stay inside and work on the book. My three kids and I also enjoyed an old-time winter holiday in the historic house.

I worked on the book through the remainder of the winter, through the summer and into the fall of 1996. During that summer, I did some part-time flight instructing for a flight school at the local airport in Michigan City, Indiana. I was instructing in an old **Cessna 172** that had seen better days. I had picked up the two students I was flying with through a little publicity I had generated in the local newspaper.

Before the 1996 winter-holiday season began, I happened upon a classified ad in Aviation Week & Space Technology Magazine that announced the need for a Challenger simulator instructor in Tucson, Arizona. The job appealed to me for three reasons: I had always wanted to qualify as a jet-simulator instructor, I had always wanted to live in Tucson and I was considering getting back into corporate flying. And besides, the prospect of another cold winter in the lake home was loosing its luster.

I picked up the phone and called about the opening. The next day, I was on an airplane headed for the FlightSafety International training center in Houston for an interview. I was hired on the spot and wasted no time getting back to Chicago to pack my things. I reported back to FlightSafety-Houston a few days later for my initial training on the Challenger.

The Challenger is one of the larger corporate jets. Therefore, captains flying Challengers for corporations commanded very respectable compensation. This was just the type of corporate-flying job I was interested in. Also, the Challenger was as complex as an airliner and had performance to match. From my limited experience flying AAR's Challenger, I knew I would be right at home on the airplane. My training began with a one-month Challenger initial-type-rating program. The program qualified me to earn my FAA Challenger type rating which made me eligible to fly as a Challenger captain. The training was interesting and on a par with major-airline training.

I then moved on to learning how to be a FlightSafety International Challenger-simulator instructor. This took an additional three months and included classroom training, a checkout as a Challenger-simulator operator and a lot of mentoring by experienced FlightSafety Challenger instructors. When my training was completed in Houston, I held a Challenger type rating (a very valuable asset) and I was ready to start instructing corporate pilots on the intricacies of flying the Challenger. This was a double whammy for me. I had just crossed two airmanship-development goals off my bucket list.

I then moved on to Tucson in early 1997 to begin training corporate pilots and enjoying life in the desert as a semi-retired airman. When I arrived at the

FlightSafety Tucson training center, the new Challenger simulator had just been installed in preparation for launching the Challenger-training program at the Tucson center. I was immediately assigned two tasks related to the launch of the program. I helped develop the training program itself, including training materials, and I worked with a couple of the other Challenger instructors on the simulator-acceptance testing that had to be completed before we could start training on it. I thoroughly enjoyed both projects. The training-program-development work gave me the opportunity to polish my PowerPoint skills and the simulator-acceptance work taught me a lot more about **Level D simulators**.

The following couple of years in Tucson were very pleasant. My FlightSafety schedule was modest (usually four five-hour days a week), the pay was adequate and I had plenty of time to explore the desert southwest. An added bonus was that I was making a lot of good contacts in the corporate-pilot world. This was a prerequisite to securing a good position after I left FlightSafety.

I had several interesting additional assignments while working for FlightSafety. I created and delivered the recurrent cockpit-resource-management training program for the Tucson FlightSafety instructors, and I created and delivered several "enrichment" classes for FlightSafety clients. I thoroughly enjoyed the simulator training and overall had a very good time in Tucson. My only major complaint was that I was only allowed to go out and actually fly a real Challenger once a year. And when I did get to fly in the real world, the clients that I was flying for generally only allowed me to perform copilot duties with no chance to fly the airplane. This shortcoming was counterbalanced with free, unlimited time in the Challenger simulator. I took full advantage of this perk.

After my first year of instructing in the Challenger program, FlightSafety nominated me to be an FAA-approved Training Center Evaluator for the Tucson Challenger program. This is essentially a check-airman authorization. After completing the FAA training for the authorization and passing the simulator checkride with an FAA inspector, I was put to use giving FlightSafety Challenger clients simulator checkrides. The FAA authorization allowed me to issue new Airline Transport Pilot and Commercial Pilot certificates with Challenger type ratings and to give charter pilots their mandatory six-month checkrides. Once again, I fully enjoyed the work and my airmanship was taken to a whole new level. Flying airplanes is one thing, instructing in them is another, but deciding whether or not a pilot is qualified to do a certain type of flying requires a whole new level of airmanship knowledge and skill.

I also learned a lot about how corporate pilots pursue airmanship excellence. In general, it was a much less structured approach than I was used to at the airlines. I found that the vast majority of the corporate pilots I worked with expended only the minimum required amount of effort to maintain their airmanship qualifications. I proposed a program to FlightSafety management that was designed to cure this problem, but I couldn't get any traction for it for various political reasons.

By early 1999, my work at FlightSafety was becoming rather redundant. I therefore decided to look for a corporate-flying job. It wasn't unusual for FlightSafety instructors to cycle in-and-out of the company, alternating stints as a corporate pilot with periods of instructing for FlightSafety. However, I didn't want to start a job search while still employed by the company. Therefore, I resigned from FlightSafety to start the job search, do some Challenger-contract flying and go back to work on the book I had put on hold to punch my ticket as a simulator instructor.

I flew a few contract trips on the Challenger as captain and as copilot. The most memorable was the trip I flew down to, and around, South America as a Challenger captain. I had flown to, from and within Europe on AAR's Challenger as a copilot, but this was my first major international trip as a corporate-jet captain. I had a good copilot and the help of an international flight-support organization — Universal Aviation and Weather. The flight went smoothly although there were several airmanship challenges that I had not faced before.

Shortly after leaving FlightSafety, I was approached by a group of FlightSafety instructors who asked me to help them to effect a culture change among the instructors in the Tucson center. They wanted to form a much-needed union. I ended up consulting to them, flying contract trips and working on the book for about three months before I was offered a job as the chief pilot of Cedarwood Construction's corporate flight department. Cedarwood is based in Akron, Ohio. They didn't have a corporate flight department when I accepted the job. That was one of the reasons I accepted the job offer. I knew I would enjoy the challenge of setting up a new corporate flight department.

The company had just purchased a **Cessna 441 Conquest II** turboprop (jet engines turning propellers). I would have preferred to fly a corporate jet (preferably a Challenger), but I had never flown a turboprop and I wanted to check that particular aeronautical-experience item off my bucket list. After completing Conquest II training at FlightSafety's Wichita training center in mid-1999, I went to Grand Junction, Colorado to oversee the refurbishment of the Conquest II. When the airplane was ready to go, I continued on to Akron to set up the flight department.

The first six months in Akron were rather hectic. I had to find a place to live and set up housekeeping while putting all the elements in place that a corporate flight department needs. While this was going on, I was flying the Conquest II quite a bit. It was, after all, a new tool for Cedarwood's executives and they all wanted to use the airplane for their regional travel. In addition, the owner of the company kept me busy flying him and his family to wherever his yacht happened to be tied up at the time. That was usually eastern-coastal resort areas in the summer, and Florida or the Caribbean in the winter.

I enjoyed running the flight department and the flying. I got to fly into many places I had not been to before. And turboprop flying requires some additional airmanship knowledge and skills that I had not acquired in the other types of flying I had done. I also had to relearn how to do all my own flight planning and dispatching because these skills had atrophied during my years flying for United. At the airline, I had a flight-dispatch team to take care of that for me.

I did have a copilot (a young woman) who I had been training, seasoning and mentoring for a few years. This was a big help. A single-pilot operation in the type of flying we were doing was much more difficult to manage safely than a two-pilot operation. Flying the Conquest II required me to go to recurrent training once a year at FlightSafety International's Wichita Training Center. I always enjoyed the ground and simulator training I received at FlightSafety. However, I did not enjoy flying from Akron to Wichita and back on the airlines. By this time, airline travel had become a real chore. And, each time I got on an airliner, I thought about how fatigued and stressed the guys upfront were. On my return airline trip from recurrent training in 2000, I vowed to never set foot on an airliner again. To date (2013), I have kept that vow.

2000s

By September of 2001, I had accumulated over 1,200 hours on the Conquest II and had wrung most of the airmanship lessons I was going to out of turboprop flying. The travel schedule had been rather hectic and I was getting tired of living on the road again even though my layovers were mostly in resort areas and I enjoyed the company of my copilot. The Conquest II was about a thirty-year-old airplane at this point and it was beginning to show the strain of heavy usage. This upped my flight-department-management workload considerably. I was spending most of the days I had in town at the maintenance hangar and in the flight-department office. I was definitely beginning to have thoughts of moving on.

On the morning of September 11, 2001, I took off from Akron-Canton Regional Airport around 8 AM EDT in the Conquest II with my copilot and six passengers. We were headed for Pocono Mountains Municipal Airport in far-eastern Pennsylvania. The airport is located about seventy nautical miles northwest of New York City. It was a beautiful early fall morning and the skies were reported clear all the way to the east coast. Our flightplan called for just under one-hour flying time.

When Akron Departure Control handed us off to Cleveland Center, the controller asked us if we would be able to fly lower than our filed altitude and maintain visual flight rules (VFR). Since it was such a nice morning and flying a little lower wouldn't appreciably affect our fuel burn, I told him we'd be able to do that and asked him what the problem was. I was expecting to hear about a radar-coverage outage. He told me that New York Center was having some kind of a problem and told me to standby. A couple of minutes later, he told us to continue our climb to Flight Level 230 (our filed altitude) and handed us off to New York Center.

When I checked in with New York Center, the controller acknowledged my check-in with a routine response and cleared us to climb to FL230 and proceed direct to Pocono Mountains Airport. It seemed that whatever problem New York Center had been experiencing was now solved. However, as we were about fifty miles west of Dubois, Pennsylvania, New York Center suggested that I land at Dubois Regional Airport. I asked the controller if he was suggesting that I land at Dubois or ordering me to land. He responded that it was a suggestion. When I asked him for more information, he told me he didn't know what was going on but he was suggesting that I land.

Since it was a clear day and I had literally hundreds of airports around me where I could safely land the Conquest II, I decided to continue on to Mt. Pocono. Our six passengers had an important meeting scheduled there for 9:30 AM. I passed my decision along to the New York Center controller we were working with and he responded casually to my message. He re-cleared us directly to Pocono Mountains Airport.

For the next few minutes, I assessed our situation and tried to figure out what was going on. I had never received a suggestion from Air Traffic Control to modify my flightplan without them giving me a good reason to do so. At this point, I was suspecting it was some kind of system problem. I did not think of this as being a national-emergency situation since I had been trained in my airline career to expect a code word to be transmitted by Air Traffic Control in that event. Since that code word had not been transmitted and the controller was not ordering me to land, I decided to continue on to Mt. Pocono.

Chapter 1: My Pursuit of More Enjoyable Personal Flying

A few minutes later, the controller handed us off to another New York Center controller and wished us a pleasant flight. The next controller cleared us for our descent into Pocono Mountains Airport. This was about the time American Flight 11 was flown into the North Tower of the World Trade Center, but we had no way of knowing that in the airplane.

As we approached Pocono Mountains Airport just before 9 AM, New York Center told me that we would not be allowed to takeoff again after we landed. When I asked the controller for more information, he told me he didn't know what was going on and he didn't know how long we would be held on the ground. I was beginning to think that the problem was much more than a system glitch and I didn't want to strand my six executive passengers in Mt. Pocono for an indefinite time. So, I requested a clearance to return to Akron-Canton Regional Airport. This request threw the controller for a loop and he told me to standby while he figured out what to do with us. We continued our descent and our direct course to Pocono Mountains Airport. A couple of minutes later, the controller gave us a clearance direct to Akron-Canton Regional Airport and told us to climb to 17,000 feet. We reversed course, started to climb and headed back home.

As we approached Dubois, Pennsylvania westbound, the controller once again suggested that we land. I told the controller that if he weren't ordering me to land at Dubois, we would continue on to Akron-Canton. He reaffirmed our clearance to continue on. The remainder of our flight home was uneventful and we landed back home around 10 AM. We didn't find out what was really going on until we walked into the lobby of the aviation-support facility at Akron-Canton and saw replays of the New York terrorist attacks on the TV in the lobby.

A few days later as more details of the attacks on 9/11 saw the light of day, I learned that hijacked Untied Flight 93 had departed Newark International Airport not long after we had taken off from Akron-Canton Regional Airport. Its flightpath went from Newark westward over central Pennsylvania before turning around in the Cleveland area and heading for Washington, DC. It crashed near Shanksville, Pennsylvania just after 10 AM. I was astounded to realize that we had been sharing airspace with Flight 93 for most of our flight to Mt. Pocono and back.

It now became clear to me why Air Traffic Control was suggesting that I land. However, I'm still perplexed to this day as to why they didn't order me to land or at least drop a few hints as to what was going on. All my training over the years had prepared me for this type of order if there was a national emergency that necessitated getting all the airplanes in the skies over the U.S. on the ground as quickly as possible. I guess we were just lucky that the

83

fighters chasing down Flight 93 didn't mistake us for the high-jacked airplane and shoot us out of the sky.

As most of us in America did after 9/11, I reassessed what I was doing relative to what I wanted to do and realized that although I had been having a pretty good time flying the Conquest II, I really had had enough of corporate flying for the time being. Adding to the reasons to move on was the fact that the owner of the Conquest II was talking about upgrading to a small corporate jet. This prospect was somewhat enticing, but it came with a requirement to do even more flying than I had been doing. I had also been thinking a lot over the past summer about resurrecting the "alternate-airline" business plan that I had been working on before going to work for FlightSafety.

In October of 2001, I notified the CEO of Cedarwood that I'd be leaving the company as soon as we could find a suitable replacement chief pilot. I continued to fly the Conquest II occasionally for Cedarwood and flew a few contract trips in a Challenger that was based in Cincinnati, Ohio. However, my focus was on the new venture.

In late 2001, I set up a not-for-profit trust named **The Aerospace Trust**. The Aerospace Trust's purpose is aerospace research, development and education. It is dedicated to the preservation of our aerospace heritage and to fostering the conditions that will maintain progress towards the aerospace industry's traditional goals of "Higher, Faster and Farther". For various reasons, I planned to use The Aerospace Trust as the vehicle for launching the "alternate airline" business plan.

In 2002, while still living in Akron, I concentrated on getting the business plan for the network of scheduled private-jet flights ready for funding. By the end of 2002, the plan was shaping up and I published my first book, *False Security: The Real Story About Airline Safety*. The book was part of the business plan in that it laid out what was really going on at the airlines and how a savvy air traveler could avoid all the risk and pain of airline travel by flying in private aircraft. It was designed to be the authentic foundation story for the new venture. The Aerospace Trust Press published the book in early 2003.

In mid-2003, I moved back to the Chicago area in preparation for launching what I called **AvWorld FliteMatrix**. The plan was similar to the Cloud 9 Air Travelers Club plan, but it included several important enhancements that I thought would make it even better. The primary change was the reordering of the roll out of the offered services. This time, I planned on launching the network of private-jet flights first and then I'd follow that up with teaching people how to fly their own private aircraft.

By mid-2004, the business plan was funded through the market-test phase with a promise from the lead investor to fully fund the plan if we got a good result from the testing. We launched the market test with a publicity campaign that generated leads. I then called on the qualified leads and invited them to give me a $2,500 deposit on an AvWorld FliteMatrix membership. I had a 100% close rate on the qualified leads with one sales call. And I was even able to get referrals from the newly signed-up members that resulted in more deposits. Small though it was, I considered that market test to be a resounding success.

However, when I went back to the lead investor, he declined to fund the marketing plan and wouldn't tell me why. I later found out that there were internal troubles in his company that he wouldn't tell me about that later resulted in a breakup of his company. Unfortunately, the other investors who had professed an interest in investing more in AvWorld FliteMatrix demurred because they didn't know why the lead investor had backed out and they suspected there was something wrong with the deal. By this time, I had a skeleton crew in place and the meter was running. I attempted to bring in new investors, but they were all put off by the refusal of the current investors to put more money into the deal.

I kept the business afloat into 2005, but we hadn't actually run any flights yet. However, we had set up the support-and-control system for the flight operation and we had secured authorization from the U.S. Department of Transportation to operate as a Public Charter Operator. We also had all of the marketing and membership-support materials and systems in place and we had made a few significant improvements to the plan. But there comes a time when a go/no-go decision must be made. By late 2005 I had become convinced that additional financing was not forthcoming with the situation we were currently in. I reluctantly shut the operation down.

After a short break, I began to rework the plan and then relaunch the business under a new name and with a new set of players as soon as I could. I didn't want to give up on the plan because I was totally convinced that the basic plan itself was sound, and that I could raise the necessary funding if I approached the problem in a different way. Fueling my enthusiasm for the plan was the fact that the AvWorld FliteMatrix members who had given me their $2,500 deposits refused to let me give them back to them. It actually took me a couple more years to get them to take it back because they wanted to secure what they believed to be a hot item and the bragging rights to being in on the ground floor on the next big thing in private air travel.

By early 2007, I had a reworked plan and a new management team ready to go. Unfortunately, by this time the run up to the Great Recession had started and I found very little interest among backers in an investment in a start up at that particular point in time. I very reluctantly put the business plan on the shelf. I decided that there was no use in trying to resurrect the AvWorld business plan until the recession ended and the business cycle started back up.

I decided to retire again. I dove back into completing the book I had started back in 1996 and spent a lot of time catching my breath after the frustrating ordeal I had just been through. I also joined a flying club at DuPage Airport to keep my hand in the flying game. I wanted aircraft that I could fly myself for my personal transportation since I had sworn off traveling on the airlines back in 2000. I also wanted to do some flight instructing.

I surveyed all of the flying clubs in the Chicago area and chose the one at DuPage Airport as the "first among equals" of the flying clubs in the area. It didn't have many of the amenities I was looking for, its aircraft fleet was old, its maintenance was somewhat lax, its flight-training program was unstructured and its aviation culture left a lot to be desired. But it was the best flying club in the area and I had a strong desire to get back into the sky. So, I joined up and started flying.

After a few months, however, as I got a closer look at the operation, I decided that the negatives of the club far out weighed the benefits and I resigned. This experience caused me to rethink the flying-club element of the business plan that was gathering dust on my shelves. It also confirmed many of my beliefs about how a flying club and flight school should be run.

During my "retirement", I took on an occasional consulting job. One of the more interesting projects I worked on was for a young entrepreneur who wanted to copy a new business model that was catching everyone's attention at the time. The company that had recently launched the new business model that my client wanted to copy was called DayJet.

DayJet held itself out as an "air-taxi", meaning you could be in any city within its operating area and summon a personal jet to take you to another city in its service area. The service was not new, but DayJet's pricing was. You could buy a single seat on a private jet rather than pay for the whole airplane as you would have to do in a traditional aircraft-charter arrangement. However, the really big advantage was that you didn't have to pay for the airplane to fly to your pick-up airport and then back to home base after dropping you off as is normally the case for a charter flight of this nature.

DayJet's Founder and CEO, Ed Iacobucci, was a private pilot who had reportedly made a fortune with Citrix Systems (a software company) that he founded and later sold. At the time I was retained by my client, DayJet had just gotten off the ground. Its fleet consisted of three recently certified **Eclipse 500** very-light jets (VLJs), and it had placed orders for 239 more of the revolutionary airplanes. Iacobucci claimed that he had created a proprietary software application that would allow him to succeed where many had failed in the past.

Many, many aviation entrepreneurs have tried to crack the air-taxi nut over the years, but they always bump into the problem of "empty legs" that drive up the overall system operating costs many fold. And the operating costs of the system have to be covered by selling single seats on revenue-generating flights. Iacobucci claimed that he would be successful where others had failed because of the reduced operating cost of the Eclipse 500 and his magic "black-box" software solution. This was apparently a good story because DayJet had raised $50-million in private-equity financing to launch the company.

I have looked at ways to crack this market several times over the years and I had come to the conclusion that it could work only if you operated very-efficient aircraft like the Eclipse 500, you kept your service area small enough and you had a customer base of several thousand people. So, at the time I was retained by the client, I held a very skeptical view of DayJet's chances of success. I told the client about my biases, and that I would agree to undertake the project only if he was willing to accept a final report that might determine that the business model was not feasible.

He agreed and I began a comprehensive and detailed analysis of the viability of starting a company modeled along the lines of DayJet. In the end, I couldn't figure out a way to make the model work without spending a vast sum of money on marketing and less-than-breakeven operations while the requisite customer base was assembled. After I reported my findings to my client, he decided to wait to see if DayJet would be successful. DayJet went out of business a short time later. This exercise confirmed my thinking that if you wanted to start a company to sell single-seats on private-jet flights, you have to do it on a scheduled basis.

In mid-2007, I came across a connection to Capt. John Rolfing who was one of my early flight instructors. He was now a retired TWA captain and he had a private airplane in a hangar at Chicago-Aurora Municipal Airport. I visited John at his hangar and he introduced me to his son. John's son was running **GLASS Simulator Center** that was based on the airport. When John's son learned about my experience as a simulator instructor, he offered me a part-time job instructing at the center. I happily accepted his offer, eager to get back into simulator instructing.

GLASS caters to pilots of high-performance single-and-twin-engine private aircraft like the King Air, Cessna 421, Beech Baron and **Beech Bonanza**. GLASS's primary market niche is in providing insurance-company-mandated initial and recurrent training on these types of aircraft. The insurance companies either won't insure pilots without the training, or the premium-

87

increases they impose are greater than the cost of the training. So, the pilots coming through GLASS, with the exception of a few professional pilots, were personal flyers who owned and flew their own airplanes for their business-and-personal transportation needs. I provided classroom and simulator training to them. We used a **Frasca 142 TruFlite** simulator for the simulator-training sessions.

For the initial training, I would spend three full days with the client. We'd normally do our ground-school work in the morning and fly the simulator in the afternoon. This gave me ample time to go over all of the classroom training and simulator scenarios required by the training syllabus. And even though the training objectives for the program did not include the passing of a checkride, I was able to do a pretty thorough evaluation of the client's airmanship knowledge and skills and how they approached airmanship in general.

I instructed part time at GLASS from mid-2007 until the end of 2010 when its business drastically contracted because the insurance underwriters, due to intense competition, were dropping their requirement for their insureds to complete the training. During the time I was with GLASS, I developed a Cessna 421 initial-and-recurrent training program and I worked with around eighty clients. This was a significant sample of a fairly broad cross section of personal flyers.

By late 2009, I was becoming very concerned about the safety of personal flyers and the people who were flying with them. I believed that the pilots I had worked with at GLASS were certainly representative of most, if not all, of the personal flyers in the U.S. who were not also professional pilots. My concern stemmed from the simple fact that none of the personal flyers that I worked with were performing at an airmanship level that would ensure that they could handle the cockpit-task loads, with an adequate margin of safety, that they might encounter in the real world. In fact, there were many instances in the simulator where the client actually crashed when a task loading similar to what they could easily encounter in an abnormal or emergency situation exceeded their airmanship capabilities.

I offered all the help I could while they were with me, but there simply wasn't enough time in the program to provide the training that would be necessary to bring them up to a safe level. And even though I counseled the clients to do more proficiency training more often, it always seemed to fall on deaf ears. I soon realized that these pilots were deluding themselves with two of aviation's most-dangerous beliefs: "It won't happen to me" and "I can do that because I've done it before". Thinking that "bad things only happen to other people" is definitely what I call "False Security". False Security is just as good as real security right up to the time it's tested. Then it's worse than

Chapter 1: My Pursuit of More Enjoyable Personal Flying

worthless. And just because you were able to, for example, shoot a safe instrument approach down to FAA minimums when you completed your instrument-rating training, it doesn't necessarily mean that you can do it at a later date unless you've been regularly shooting them and undergoing frequent recurrent training.

Almost all of the personal flyers I worked with at GLASS also had a bad case of complacency. This complacency usually came from flying the same personal airplane for several years, a false sense of security and not being regularly challenged to perform at a higher airmanship level. After all, it is a known human-factors fact that we all have a tendency to do only what we have to do to be legal and minimally qualified to do something we want to do. I was seeing this phenomenon close up. The only training these pilots got was the once-a-year insurance-mandated recurrent training. Based on their observed performance, this was woefully inadequate.

The situation is even worse when it comes to personal flyers who are not flying high-performance airplanes and are therefore not subject to an insurance mandate to take recurrent training once a year. These personal flyers only have to fly with a flight instructor once every two years for one hour to remain legal to fly their personal aircraft. Based on my observations of personal fliers over the years, I believe most personal flyers should undergo a ground-school, simulator and airplane recurrent-training session every ninety days to achieve and maintain the level of airmanship they need to safely fly their airplanes, even simple ones.

Another major factor that I identified while working at GLASS is the fact that virtually none of the personal flyers I have worked with, or have known, have a personal airmanship-development plan. And without a personal airmanship-development plan, a personal flyer can't acquire and maintain the airmanship knowledge and skills needed to be a safe and proficient pilot. In other words, very few of the pilots I worked with at GLASS were aggressively and effectively pursuing airmanship excellence, and it showed.

To better understand the problem, I researched the causes and rates of general-aviation accidents. I was, of course, generally aware of this information through my normal monitoring of the aviation industry, but I hadn't focused on the details of personal-flying accidents for some time. This was because the causes of general-aviation accidents have been virtually unchanged for more that four decades. I turned first to the **2009 Nall Report** published by the **Air Safety Institute**. It is based on **FAA accident data** and **National Transportation Safety Board** accident analyses. The Nall Report has been published annually for over twenty years and it is considered to be the "go-to" source for reliable general-aviation-industry safety data.

The 2009 Nall Report includes accident and trend data for years 1999 through 2008. The report shows that there were 927 personal flying accidents in 2008 (73% of all general-aviation accidents) with 186 being fatal accidents (77% of all fatal general-aviation accidents) with 319 fatalities (74% of all general-aviation fatalities). The Nall Report breaks down the causes of these accidents into seven categories: fuel management, weather, takeoff and climb, maneuvering, descent and approach, landing and other. All of the things a pilot can do to reach a bad ending fall into one of these categories. These categories have been used by the aviation industry for over three decades. We have known for a long time how pilots kill themselves in airplanes, and we have known just about as long how to prevent ourselves from doing that.

Professional flight organizations have learned how to train, monitor and support aviators so they don't make these mistakes, and if they do, how to recover from them. Personal flyers, lacking the motivation and support structure that would enable them to learn how to stop making these errors, keep right on making them. This conclusion is borne out by the 2009 accident rate per 100,000 hours flown. For the airlines it was 0.17 and for personal flyers it was 6.6. This means that the accident rate for personal flyers is about thirty-nine times worse than the accident rate for airline pilots.

I was well aware of ongoing FAA and aviation-industry efforts to educate private flyers about excellent airmanship. Every month, several aviation publications carry a multitude of articles that can help an airman improve his level of airmanship. The **FAA Wings Program** runs seminars all over the country promoting aviation safety and offering valuable aviation-safety-related programs. The Aircraft Owners and Pilots Association (AOPA) also produces and makes available at no cost a wide variety of excellent aviation-safety seminars, Webinars and online-training courses. The Experimental Aircraft Association (EAA) does an excellent job of producing and disseminating safety information that is related to experimental, home-built aircraft. The overall airmanship-related resources available to personal flyers are voluminous and easily accessible.

The problem is that this robust effort to educate personal flyers on how to become better airmen has been going on for well over twenty years with little-to-no effect on the personal-flying accident rate. In 1992, the general aviation accident rate was 8.5 accidents per 100,000 hours of flying time. In 2010, it was 7.2. The rate fluctuated above and below the resulting trend line over that period (1992 – 2010). Clearly, it isn't rational to assume that continuing to rely on education alone is going to make personal flying safer any time soon.

Another inconsistency is the puzzling fact that although thousands of modern aircraft with advanced safety features have been coming into the personal-flying fleet, the accident rate for personal flyers flying the modern aircraft is about the same as that for pilots flying older aircraft. The Cirrus SR20 was the first of these modern aircraft. It entered the market in 1999. These modern aircraft are equipped with safety devices like advanced flight-automation systems, cockpit-computer displays (glass cockpits), traffic alerting systems, GPS navigators, terrain-collision-avoidance systems, satellite XM in-cockpit weather, satellite phones, whole-aircraft parachutes, air bags and stronger, more-crashworthy airframes.

It has been demonstrated in the professional-flying world that the use of the most-modern safety devices significantly reduces accident rates. So, why aren't the personal-flying accident rates for modern aircraft much lower than that for non-modern aircraft? I believe that I'm safe in saying that the consensus among the experts who have looked at this dichotomy is that the root cause of this problem is the lack of proper initial and recurrent training on how to properly operate this safety equipment.

This dearth of training is coupled with a deficiency of proficiency in operating glass cockpits. When the pilots flying these modern aircraft encounter technology overload and the demands on their airmanship capabilities exceed their current level of airmanship, they are just as likely to end up as an accident statistic as somebody who is also lacking in his or her airmanship skills and flying a thirty-year-old airplane. However, most experts agree that good initial and recurrent training, including simulation, will solve this problem just as it has done for the airlines. This training is available to pilots of modern personal aircraft. Unfortunately, very few of them avail themselves of these life-saving airmanship-development opportunities.

After a lot more research and considerable thought, I could see at least a vague outline for the solutions to the problems of: motivating personal flyers to "fly right", training personal flyers to be fully qualified to do the kind of flying they want to do and to providing personal flyers with the support they need to effectively pursue airmanship excellence. I knew there was a lot more work to do to fill in the outline. However, based on my experience, I believed that a comprehensive solution to these problems could be created and implemented.

In late winter 2009, I started bouncing my thinking off other senior aviators whose judgment I trusted. I talked with several aviators and aviation-business professionals about the problems faced by the personal-flying segment of the general-aviation industry and those faced by the personal flyers in that segment. I apologize in advance for leaving out the names of some of these advisors. However I want to mention a few of them and how they've contributed to coming up with the solutions presented in this book.

David Greenberg was the first trusted advisor that I talked with. Dave had advised me on the relaunch of the AvWorld FliteMatrix plan. Dave is a little senior to me. He started his aviation career while in college by earning his private-pilot certificate from a local flight school. After college, he moved on to the U.S. Air Force where he served as a **T-37** and T-38 instructor pilot. After his tour in the Air Force, he hired on with Delta Airlines and moved up through the ranks there to become Vice President of Flight Operations.

Dave took early retirement from Delta in the late 1990s and set up his own aviation-consulting business. He was hired by **Korean Airlines** to straighten out a serious airmanship problem the airline was experiencing. Dave completed that assignment successfully and then moved on to founding and running **Cargo360**, a Boeing 747 cargo airline. When Cargo360 was merged with Southern Air in 2008, Dave went back to running his consulting business. He continues to be a trusted advisor and is a current member of the Center For Airmanship Excellence Board of Advisors. He has made several important contributions to my thinking on the solutions to the personal-flying problems cited above.

For example, he made significant input by suggesting to me that I give what I had been talking about "a name". Up to that point, our discussions had centered on the nuts and bolts of how we could take the best practices of airline and military pilots and implement them in a way that personal flyers could embrace. The solution we identified is comprehensive and therefore complex. Dave saw the need to "put a handle" on all this detail. I agreed with him and started thinking about how to put a framework around the concept.

My inquiries and ruminations led me to conclude that "airmanship" was the broad name we could put on it. This decision led me to research airmanship in detail. I soon found a book titled ***Redefining Airmanship***. Although it was published in 1997, somehow it hadn't appeared on my radar screen until 2010. Col./Dr. **Tony Kern** wrote it. He is a retired USAF Command Pilot with a strong background in pilot training, human factors and education,. He has written five books on airmanship. I highly recommend all of them to any aviator who is, or who is thinking about, pursuing airmanship excellence.

I couldn't put the book down. In it, Kern covers the origins, essence, and principles of airmanship. He illustrates his key points with solid data derived from military and civilian research into airmanship, and case studies of accidents that were caused by deficiencies in airmanship. He also includes his "airmanship model" that I now call "**The Kern Airmanship Model**".

When I finished the book, I was astonished to note that I didn't find anything in it that I disagreed with. This was surprising because I usually find at least one point made in aviation-related publications that I do not agree with.

It wasn't long before I made the decision to use the principles and precepts laid out in "Redefining Airmanship" to form the framework for what I wanted to talk about with the personal-flying community.

I had been attending FAA Wings Program safety seminars since early 2009. Regular attendance at aviation-safety seminars is a key element of my Personal Airmanship Development Plan. The FAA Wings Program offers a wealth of airmanship-development knowledge and know-how. Also, attendance at the seminars afforded me the opportunity to see what other personal flyers were thinking and to bounce my ideas about airmanship off some of them.

I met **Dave Shadle** fairly soon after I started attending the FAA Wings Program seminars. We started talking about what was wrong with the safety culture within which virtually all of the personal flyers in the U.S. operate. It wasn't long before it was obvious that Dave and I were in agreement on the major issues facing personal flyers. This was encouraging to me because I greatly respect Dave's expertise and knowledge.

Dave Shadle started flying in May of 1958 in his hometown of Clinton, Iowa and soloed an Aeronca Champ on his 16th birthday. He went on to earn his Flight Instructor certificate in 1969 and instructed in Davenport, Iowa for about five years. In January of 1972, Dave became an accident prevention counselor and was active in the FAA's Accident Prevention Program, initiating and participating in several clinics in the Quad Cities area. He was appointed as a Designated Pilot Examiner for the FAA's Des Moines office in 1973. He was named Flight Instructor of the Year in the FAA Des Moines Flight Standards District Office in 1974.

Dave then went on to be a corporate pilot in Iowa and Illinois, flying primarily Beechcraft King Airs and **Falcon 20s** before returning to the Quad Cities to manage the flight department at **Elliott Aviation**. He was the Chief Flight Instructor, Chief Pilot, Director of Operations and Check Airman for Elliott's operation and was appointed a Designated Pilot Examiner in the DuPage Flight Standards District Office in 1977. In 1987, Dave moved to Decatur, Illinois and was a pilot for Archer Daniels Midland, flying a **King Air 200** and **Falcon 10s**.

He joined the FAA in 1990 as an Aviation Safety Inspector (operations) in the DuPage Flight Standards District Office. During his time with the FAA, he traveled throughout the country as the national resource for the **King Air 300/350**, conducting flight checks and type rating tests. He also did regional resource work for the FAA in Cessna Citations. After retiring from the FAA in 2001, Dave again became an FAA Designated Pilot Examiner and did some aviation consulting work. He has started five private-aircraft-charter operations

93

and expanded five more. He has written numerous operations manuals for charter operators.

In November of 2001, he also became a charter member of the Great Lakes Wings of Angels Flight America, which is the largest group of volunteer pilots in the country providing transportation to people in need of specialized medical care. In May of 2010, Dave received the FAA's Wright Brothers Master Pilot Award. This prestigious award recognizes pilots who have demonstrated professionalism, skill and aviation expertise by maintaining safe operations for fifty or more years.

Dave has accumulated over 12,000 hours of flight time, and given nearly 4,800 hours of flight instruction. He holds an Airline Transport Pilot certificate with both single- and multi-engine ratings and is type rated on the Cessna Citation, Beech King Air 300/350 and the **Beech 1900 Airliner**. He is also a gold seal Certified Flight Instructor with a single- and multi-engine airplane and instrument-airplane ratings, as well as a Ground Instructor with Advanced and Instrument ratings.

When I met Dave in late 2009, he was a Lead Representative for the **FAA Safety Team** (FAASTeam). The FAASTeam's mission is, "To improve the Nation's aviation safety record by conveying safety principles and practices through training, outreach, and education. At the same time, FAASTeam Managers and Program Managers will establish meaningful aviation industry alliances and encourage continual growth of a positive safety culture within the aviation community. The FAA trains and fields volunteers who actively promote aviation safety within the personal-flying community."

The FAA's primary vehicle for fulfilling its mission is the FAA Wings Pilot Proficiency Program (WINGS). "The objective of the WINGS Program is to address the primary accident causal factors that continue to plague the general aviation community." The WINGS Pilot Proficiency Program is based on the premise that pilots who maintain currency and proficiency in the basics of flight will enjoy a safer and more-stress-free flying experience. The WINGS Program consists of learning activities and flight tasks selected to address the documented causal factors of aircraft accidents.

2010s

Since my personal objectives aligned very nicely with those of the WINGS Program, in early 2010, at Dave Shadle's invitation, I volunteered to serve as a FAASTeam representative. FAASTeam representatives organize and participate in FAASTeam events, initiate action to correct conditions that may be hazardous to persons or aircraft in flight or on the ground and counsel airmen. Dave Shadle introduced me to the two FAA safety inspectors who were responsible for the Wings Program in the Chicago area at that time—Sam Heiter and Tim Sokol. Sam is an operations inspector and Tim is an airworthiness inspector. I was very impressed with their knowledge of, and enthusiasm for, the FAA Wings Program. Both of them were very helpful in training me to be a Wings Program volunteer representative, and they both encouraged me to explore my ideas concerning airmanship as it relates to personal flyers.

After receiving my initial FAASTeam-representative training from Sam and Tim, the first thing that I did was to organize a series of seminars that examined Kern's book "Redefining Airmanship". I wrote and produced PowerPoint presentations to provide a framework for each seminar and a basis for discussion. I delivered my first WINGS Program seminar on June 14, 2010 at the Blue Skies Pilot Shop in Lake In The Hills, Illinois. It was titled "Redefining Airmanship: Your Free Personal Airmanship Development Plan". By August 17, 2010, I had conducted eleven of these WINGS Program events around the Chicago area. About two hundred pilots in total attended. Each event included an exit survey.

While I was working on the WINGS program events in early 2010, I was also writing a full business plan for what I later dubbed an Airmanship Development Support Organization (ADSO). In March of 2010, I published a thirty-eight-page business plan for the ADSO. The plan was based on a detailed financial model of the operation. I wanted to make sure that the solution I was thinking about to the two big problems facing personal flyers could stand on its own as a business proposition. I shared the business plan with the growing group of trusted advisors who had gravitated to the project.

I reviewed the feedback I received from the WINGS-seminar participants, the results of the WINGS Program-event surveys and the feedback I received from my group of trusted advisors and I came to four basic conclusions:

1. Well over half of current personal flyers would like to have a personal-airmanship-development plan.

2. A large number of personal flyers would like to fly with a professionally run flight-operations-support organization.

3. We could use "airmanship" as the framework for implementing a culture change among personal flyers.

4. An ADSO could be financially viable.

I also came to the conclusion that we needed to set up an organization to facilitate the culture change we believed was necessary to significantly improve the personal-flying accident record and to appreciably expand the number of personal flyers in the U.S. So, in early August 2010 I established the Center For Airmanship Excellence (CFAE) as a division of The Aerospace Trust. The original organization chart for the Center consisted of a Board of Advisors, an Executive Director and four committees (Training Operations, Safety and Standards, Flight Operations Support and Airmanship Outreach).

The initial members of the CFAE Board of Advisors were Capt. H. David Greenberg, Mr. David Shadle, **Capt. Arnold Quast** and me. Arnie is a captain for United Airlines. He has been flying as a professional pilot for over twenty-five years, and he is very active in both airline-pilot and personal-flying circles. Arnie started his flying career as a line-service technician and then as a primary flight instructor. He went on to complete a Bachelor of Science in Aeronautical Science degree at Embry-Riddle Aeronautical University (ERAU)—one of the most prestigious aviation universities in the world. I met Arnie at an ERAU alumni meeting, and I immediately took a liking to him because of our similar backgrounds and his "can do" attitude.

The initial Center For Airmanship Excellence committee chairs were **Capt. Bill Brand** (Training Operations), **Capt. Richard Sternal** (Safety and Standards), **Mr. John Keiper** (Flight Operations Support) and Mr. James Binder (Airmanship Outreach). Bill Brand has one of the most diverse airline-training backgrounds that I've ever come across. Bill retired from United Airlines in 2004. He started his long career in aviation in 1962 when he joined the U.S. Air Force. After leaving the Air Force, Bill entered general aviation as a flight instructor and he flew as a volunteer pilot in South America for Wings of Hope. He also served a stint as a bush pilot in Alaska. Over the years, Bill flew for, and worked in the training departments of, seven airlines. After retiring from United, he earned a Masters Degree in Aeronautical Science from ERAU with an emphasis on aviation education.

Rich Sternal started out in aviation as an aeronautical engineer for Boeing. However, it wasn't long before his passion for flying led him to a bush-pilot job in Alaska. Rich went on to become an American Airlines captain. After taking an early retirement from American, Rich got into general aviation as a flight instructor. He is currently the general manager of the GLASS simulator center.

John Keiper has been a personal flyer since the early 1970s. John served as a law-enforcement officer in Indianapolis for over seventeen years. He also has extensive experience in IT systems and flight-operations management. John founded and ran his own destination-management business. His brother is a veteran corporate pilot.

James Binder holds an FAA Private Pilot Certificate with an Instrument Rating and an FAA Dispatcher Certificate. He has served as an instructor for the Lewis University Flight Team where he earned a Bachelor of Science degree with a major in aviation administration and an MBA degree with a concentration in management-information systems. He has also worked for United Airlines as a Flight Crew Data Coordinator and for American Eagle Airlines as an Operations Agent/Load Planner. He is currently serving as a flight dispatcher for United Airlines.

CFAE was pretty well organized by the end of August 2010 and we began to hold regular meetings as well as frequent committee and advisory-board meetings. We were focused on finding a comprehensive solution to personal flying's two most-critical problems: the personal-flying accident rate and the rapidly dwindling numbers of personal flyers. We also went live with the **CFAE Website** in August 2010.

In September 2010, I launched **Capt. Dave's Hangar**, my personal Blog about airmanship. At this time, I also repackaged the "Redefining Airmanship" seminar series into the **Redefining Airmanship Wings Program Webinar Series**. CFAE and the FAASTeam jointly sponsored several of these Webinars during the last quarter of 2010. The Webinars were recorded (video) and posted on the CFAE Website in the **Airmanship Archives** section. The exit surveys from these Webinars and the feedback I received from my Blog confirmed our earlier results and encouraged me to continue to seek the comprehensive solution I was looking for.

In November 2010, I was invited to join the **Kishwaukee College** (KC) Aviation Advisory Committee. Mr. Steve Durin extended the invitation to me. Mr. Durin is the school's Aviation Program Coordinator. KC is the only junior college in northern Illinois with an aviation program and the school is very active in networking to high-school students who are interested in a career in aviation. Steve is an enthusiastic personal flyer and aircraft owner and he is very active in helping young people to realize their aviation dreams.

In early 2011, I made a personal commitment to focus more closely on the nuts-and-bolts of the plan that had been coalescing throughout 2010. By this point, I was pretty well convinced that a comprehensive plan could be developed to address the personal-flying safety and population issues that we had identified as being crucial to turning around the personal-flying segment of the general-aviation industry. But as my old grandfather used to always tell me, "the devil is in the details." My experience as an entrepreneur has taught me that you can't be sure a plan will work until you thoroughly examine all the pertinent details and quantify the financial aspects.

During the first quarter of 2011, I developed and delivered a series of FAASTeam WINGS program seminars titled "Airmanship For New Airmen". The seminars were designed to explain to student pilots, recently certificated private pilots and aspiring pilots the "right way" to approach flight training and flying as a personal flyer. I took this step because most of the research I had been doing in 2010 was focused on experienced personal flyers and I wanted to see if people who fell into one of these other categories would accept the airmanship-development message. The feedback I received from seminar attendees confirmed what I believed about recently certificated airman and potential personal flyers. That is that if folks who are interested in becoming personal flyers are honestly and comprehensively educated about "flying right", they will follow the airmanship-development path needed to reach that goal.

In July of 2011, the Center For Airmanship Excellence (CFAE) signed a Letter of Understanding (LOU) with the Federal Aviation Administration Safety Team (FAASTeam). The LOU established an arrangement between the FAA and CFAE to promote aviation safety awareness and training to all airmen. With this LOU, CFAE became a FAASTeam Industry Member.

In early August of 2011, the Center For Airmanship Excellence (CFAE) launched a division called the **National Association of High School Aviation Clubs** (NAHSAC). NAHSAC is dedicated to helping high-school students, their parents, their teachers and school administrators to establish aviation clubs in their high schools. We believe that it is important to reach out with our airmanship message to young people who may want to learn to fly. We also want to reach the parents of interested students who have themselves been harboring a desire to learn to fly.

Kishwaukee College (through Steve Durin) was the first organization to join CFAE as a cosponsor of NAHSAC. Shortly thereafter, William Brogan (the Chair of the Aviation Department at Lewis University) committed **Lewis University** as a NAHSAC cosponsor. Then, David A. NewMyer, (Chairperson and Professor of Aviation Management and Flight, College of Applied Sciences and Arts, Southern Illinois University) brought **Southern Illinois University** (SIU) onboard as a cosponsor, and Capt. Dolores Pavletic (FedEx) arranged for the Chicago chapter of **Women In Aviation International** to also cosponsor NAHSAC. Capt. Pavletic had been doing a lot of pilot-recruitment work for FedEx and she is one of the movers-and-shakers in the aviation industry's efforts to encourage more women to come into the industry.

In the summer of 2011, Jan Lebovitz of the FAA Great Lakes Region's **Aviation and Space Education Program** invited NAHSAC to work with the FAA to establish an aviation-career pipeline in the Chicago area. Before the summer was over, Jan had arranged for me to announce NAHSAC at an FAA

ACE Camp in Indianapolis, Indiana. The students and teachers who attended the ACE Camp were very interested in NAHSAC. We even garnered a little press coverage in **General Aviation News**.

In September of 2011, I met **Dr. Frank Bacon**. He had attended one of my FAA Wings Program airmanship Webinars and he called me to volunteer to help the CFAE team in any way that he could. Dr. Bacon is an accomplished, long-time personal flyer. Over the years, he has flown his own **Cessna 170**, **Piper Archer**, Beechcraft Barons and a Beechcraft King Air for his business-and-personal-mobility needs.

Dr. Bacon is the Founder and Chairman of the **Planned Innovation Institute, Inc**. He is an internationally recognized authority on new product/service development and strategic planning. Many years ago, he developed an approach to new-product/service introductions that has proven to be very effective. He calls this approach **Planned Innovation®**. It is a market-oriented, scientifically based, entrepreneurial approach to achieving higher levels of performance in new or existing organizations.

Over the years since the founding of the Planned Innovation Institute, Dr. Bacon has trained hundreds of managers in Fortune 500 companies to use his Planned Innovation process. It has been shown to be extremely effective, achieving a 97% success rate for one Fortune 50 company. Dr. Bacon joined the CFAE Board of Advisors shortly after I met him. He has been instrumental in helping me to apply his Planned Innovation process to help solve the major problems facing personal flyers.

Around this time (fall 2011), I joined the **Chicago Flight Instructors Association (CFIA)**, and I have regularly attended CFIA monthly meetings since then. The meetings are always informative and I truly enjoy seeing many of my compatriots from my United flying days and my Chicago-area general-aviation flying. Capt. Vern Jobst has been the president of the CFIA since 1995. He is a retired United Airlines pilot. Vern has been a prominent figure on the personal-flying scene for several decades. He is still qualified to fly as a Pilot In Command, and still flies regularly, at the age of eighty-two.

Capt. Jobst first soloed in 1948 as a personal flyer. Vern has accumulated over 43,000 flying hours in a professional airmanship career that includes Capital Airlines (hired in 1951) and United Airlines (Capital merged with United in 1961). Vern retired from United with over forty-years of service. His airmanship credentials are too numerous to cover in this book, but they include twenty-six large-aircraft and jet type ratings, he was the second president of the **EAA International Aerobatic Club (IAC)**, he is a flight-and-ground instructor, he flew the **EAA Spirit of St. Louis** on a U.S. tour in 1978, he has checked out in over 362 various types of aircraft and he has owned, or been a partner in, several personal-flying aircraft. Vern has set several aviation records in aircraft ranging from Piper Cubs to Boeing 747s. Vern still serves as an FAA Designated Pilot Examiner and he is very savvy about the psyche of personal flyers.

Shortly after joining the CFIA, I renewed my acquaintance with Vern over lunch. I wanted to get his input on what I was thinking about how to reinvigorate personal flying and how to make it safer. He provided me with very valuable insight into what would be needed, in his opinion, to change the personal-flying culture in the U.S. Vern has also carved time out of several CFIA meetings to allow me to make airmanship-related presentations to the CFIA membership and to solicit their feedback. Capt. Jobst has also encouraged me to keep going on my quest to find the right solutions to the major problems facing personal flying today.

In the fall of 2011, I started flight instructing part time at a Chicago-area flight school. I wanted to update my first-hand experience working with people who were thinking about learning to fly and with new student pilots. I gave introductory flights and primary flight instruction in Cessna 172s and **Diamond DA-20s**. It was about this time that I coined the term "Airmanship 2.0" to differentiate what I was calling "the right way to fly" from the way virtually all personal flyers are doing it (Airmanship 1.0). Cirrus Aircraft's use of the term "Flying 2.0" inspired me to select the term.

Throughout the remainder of 2011, I developed, produced and delivered four additional WINGS program seminar/Webinar series: **Kern Airmanship Model In Detail**, **Airmanship 2.0**, **Airmanship 2.0 Case Studies** and **Airmanship Challenge Program**. Each series was made up of several individual seminars and Webinars. They were offered multiple times to pilots in the Chicago area. These seminars and Webinars allowed me to refine and expand the CFAE airmanship-development message and to get feedback on the effectiveness of the message.

Also in 2011, we expanded the Center For Airmanship Excellence committee structure by adding the Aircraft Selection and Maintenance Committee. **John Nowicki** volunteered to chair it. John has over thirty years experience in general aviation as a personal flyer, FAA-certificated airframe-and-powerplant mechanic and aerobatic-competition pilot. I also consulted throughout 2011 with **Tim Perry** on ADSO-related personnel matters. Tim is a former USAF Navigator and a human-resources expert. I met Tim through Arnie Quast at an Embry-Riddle Aeronautical University Alumni meeting.

By the end of 2011, I had become convinced that a culture change could be effected within the personal-flying community that would, if adopted by a sufficient number of personal flyers, significantly improve the personal-flying experience. And I concluded that the new culture could attract new pilots on a scale that could turn the declining number of personal flyers into a rapidly growing personal-flyer community. One of the things that convinced me that this culture change was possible was the fact that by the end of 2011 over two-dozen current personal flyers had volunteered to help CFAE do just that.

For example, one of those volunteers was Doug Willey. After college, Doug served as a U.S. Navy surface-ship engineering officer. In 1989, he went through Navy flight training where he flew **Beech Turbo Mentor T-34Cs** and **Beech King Air T-44As**. After flight training, Doug flew the **P-3 Orion** in support of Desert Storm. After his active-duty service, he flew **Gulfstream C20Gs** for the Navy Reserves. The C20G is the military version of the very popular Gulfstream GIV corporate aircraft. Doug currently flies for United Airlines as an **Airbus A320/319** first officer. He is also a personal flyer and helicopter pilot. Doug has made several important contributions to our thinking about personal flying.

In early 2012, I committed myself to proving that a personal-flying culture change could work. However, my industry experience and knowledge led me to believe that this culture change could not be achieved merely by talking about it. Even if talking about it included continuing to expand the reach of the CFAE-WINGS program airmanship seminars and Webinars and "preaching to the choir" whenever the opportunity presented itself. I had believed for some time that the personal-flying community would not embrace the needed culture change in any meaningful way until it could be demonstrated to all the stakeholders that the culture change could actually be made. I decided to move ahead with this essential demonstration by standing up an **Airmanship Development Support Organization Demonstrator** (ADSOD) in the Chicago area.

I set about completing a new financial analysis and writing a new business plan since my thinking about the best way to stand up the ADSOD had evolved rather dramatically since I wrote the first business plan in 2010. In the first quarter of 2012, I met with potential management-team members, candidate staff, vendors, strategic partners, attorneys, accountants, the FAA and my trusted advisors. Everyone was enthusiastic about being involved with a project with industry-changing potential and they all freely offered their expertise, ideas and encouragement.

One of the more helpful advisors who informally joined our team in 2012 was **Peter Halasz** the president of Phyxius Inc. Phyxius is a dedicated group of trusted aviation professionals with wide-ranging expertise. They provide a variety of services ranging from fee-based aircraft acquisition to aircraft sales and aircraft management.

Another trusted advisor who came onboard in 2012 was **Michael Lockett**. Mike has over forty years of experience in business accounting and business-start-ups. He has applied his expertise to the many accounting and tax issues that are inherent in the launch of an ADSOD. Jim Baloun is a very experienced CPA and the owner of **CPAaviation.com**. JIm has aviation expertise in accounting, asset protection, finance, record keeping, aircraft acquisition and Taxation. He is a personal flyer and owns his own airplane. Jim helped Michael and me to better understand the tax issues related to aircraft leasebacks.

I published the new business plan in April 2012. It contained my best thinking on standing up the ADSOD. However, there was still a veritable battalion of devils (details) to deal with before we could fund and implement the plan. Therefore, I began to address each one of those details in earnest.

During the first half of 2012, I also wrote, produced and delivered several more CFAE-WINGS Program Webinars that were devoted to broadening the Centers approach to delivering its airmanship-development message. Attendance at these Webinars steadily grew as we expanded their geographical reach. By the end of 2012, over 560,000 emails had been sent to Chicago-area pilots announcing CFAE-WINGS program airmanship seminars and Webinars. Almost 2,000 pilots registered for these events and total attendance at these events was about 1,000 (approximately 10% of the pilots in the Chicago area). The feedback that I received from the attendees confirmed our earlier results.

In the spring of 2012, I was invited to join an ad hoc committee that was working with the FAA to improve the FAA WINGS Program. I was invited to join the committee by Paul Berger. Paul is the Executive Director of the **Advocates For Aviation Safety Foundation**. Paul has been working tirelessly for several years to improve the effectiveness of the FAA WINGS Program. He was the catalyst for the formation of the committee and he served as its first chairman.

On May 11, 2012, the Center For Airmanship Excellence signed an LOU with the FAASTeam that established an agreement between the FAA and the industry members of the committee to form the WINGS Industry Advisory Committee (WIAC). The WIAC was assembled to promote aviation safety by examining possible improvements to the FAA WINGS Program. So far, the committee has met via telephone-conference calls.

In January 2013, two airline pilots joined the Center For Airmanship Excellence as members of the Airmanship Outreach Committee. **Andrew Roccasalva** came onboard as the chair of the committee and **Joshua Allison** joined the committee as the Executive Director of the National Association of

High School Aviation Clubs (NAHSAC). Both of them fly for SkyWest Airlines, Joshua as a captain and Andrew as a first officer.

All in all, many highly qualified and experienced airmen have advised me on the concepts and ideas that are in this book. Some of them have joined our team to pursue the goal of introducing Airmanship 2.0 to current and potential personal flyers. I believe that more will join us as we turn the vision into a reality. At present (Fall-2013), the team is working to do just that. If you'd like to stay up-to-date on our progress, you'll find updates on the Center for Airmanship Excellence Website.

Well, that's the story of the evolution of my thinking about airmanship excellence, how to go about pursuing it and in what ways personal flying can me made more enjoyable. I've always been concerned about air-travel safety and convenience. Therefore, I've tried to contribute to the advancement of both whenever I've had the opportunity. I'm thankful that I now have the chance to help personal flyers to make their flying more enjoyable through the pursuit of airmanship excellence.

As you surely noted, the story spans more than fifty years of my professional and personal flying. When I was a kid, I dreamed about all the flying adventures I was going to have. As a senior aviator, I dream about all the flying adventures I've had. I apologize for not sharing more war stories about my experiences during the golden age of airline flying. I do have a few. However, due to the focus of this book, I feel that it is better that I save them for my memoirs. I plan on publishing them when my youngest child turns fifty. He's twenty-seven now.

I hope this chapter gave you the information you needed to be able to trust what is written in this book, and to see how the continuous pursuit of airmanship excellence makes my personal flying more enjoyable. I firmly believe that if one is going to fly oneself for recreation and/or business-and-personal mobility, it is imperative for many reasons to pursue airmanship excellence. That's flying right. And, flying right leads to more enjoyable personal flying. Won't you join me?

Chapter 2: Airmanship 2.0

> *From knowing himself and knowing his airplane so well that he can come somewhere close to touching, in his own special and solitary way, the thing that is called perfection.*
>
> —***Richard Bach***, A Gift of Wings

The suffix "2.0" usually denotes a new way of doing something that differs a great deal from the old "1.0" way. Airmanship 2.0 replaces the antiquated and less safe attitude towards personal flying of the past (what I refer to as Airmanship 1.0) with a modern personal-flying methodology that is efficient, productive and enjoyable. Airmanship 2.0 provides a support structure for your personal flying that ensures a world-class personal-flying experience. Airmanship 2.0 employs the same tried-and-true techniques, policies and procedures used by contemporary professional pilots to provide a new paradigm in personal flying for those pilots who adopt it as their approach to airmanship.

Why We Pursue Airmanship Excellence

Let's begin our discussion of Airmanship 2.0 with a close look at why we pursue airmanship excellence. The "we" that I have in mind includes me, the **Center For Airmanship Excellence** volunteers who are helping pilots in their pursuit of airmanship excellence and the thousands of professional pilots who pursue airmanship excellence on a daily basis. During my fifty-plus years as a pilot, I have known and flown with thousands of other pilots. Most of the professional pilots in this group have been, in one way or another, pursuing airmanship excellence. And all of the professional aviation-support-organization cultures that I have been a part of have held the pursuit of airmanship excellence as a core belief.

Unfortunately, most of the personal flyers (those pilots who fly airplanes for their business-and-personal-travel needs) that I know do not. I think I have a good handle on why these personal flyers aren't pursuing airmanship excellence. I'll explain later how I believe we can motivate a large percentage of them to do so.

So, why do we pursue airmanship excellence? The reasons are many and vary from pilot to pilot. However, I believe we can boil them all down to some combination of the following:

- It's required by the airmanship culture to which we belong.

- We know that by pursuing airmanship excellence our flying will be as safe as is humanly possible.

- We know that by achieving airmanship excellence, we will be recognized for our efforts and granted ever-increasing aeronautical privileges and responsibilities.

- We are self-motivated to pursue airmanship excellence by the rewards we get out of the pursuit.

- We gain the respect of our peers.

- We enjoy mastering the challenges we encounter in our pursuit of airmanship excellence.

- We have willingly accepted the solemn duty to protect our loved ones and passengers, and we know that the best way to protect them is for us to pursue airmanship excellence.

Airline and military pilots are told by their bosses that they must pursue airmanship excellence through the highly structured training, safety-management and professional-development programs that they have in place; or they will lose their jobs. Most of us who are not professional airline or military pilots don't have anyone telling us to pursue airmanship excellence. Therefore, we must truly want to do it for one or more of the above reasons.

The most important reason why I pursue airmanship excellence is because I believe that I have a sacred responsibility to my passengers and fellow crewmembers. If I take someone aloft with me, I am totally responsible for their wellbeing while they are flying with me. I have learned over the years that the best way to protect my passengers and crew is to continuously and consistently pursue airmanship excellence. The second most important reason why I pursue airmanship excellence is that I have a responsibility to my loved ones to return safely from every flight that I embark upon. My pursuit of airmanship excellence is the "insurance policy" that I use to ensure that the number of my safe landings equals the number of my takeoffs.

These two reasons should be enough to convince any pilot to pursue airmanship excellence. However, there is also another very good reason to pursue it—self-fulfillment. In my opinion, the respect that airmanship excellence compels others to bestow upon us is, in itself, more than worth the time and effort required to achieve airmanship excellence. And the self-satisfaction that comes from declaring a challenging goal and then reaching it is worth even more than that.

If you are flying, or if you intend to fly, light, general-aviation airplanes for your business-and-personal mobility needs, you should be very interested in airmanship excellence not only for the above reasons, but also from pure self-interest. That self-interest concerns your personal security and your economic wellbeing. And it concerns your future experience of personal flying in the U.S.

For example, one of the tenets of the pursuit of airmanship excellence is spreading the word about it to other personal flyers and aspiring aviators. It is my belief that this "airmanship outreach" will pay awesome dividends into our self-interest account. Those dividends include increased personal security, better economic wellbeing and an enhanced personal-flying experience.

We increase our personal security by bringing as many other pilots into the pursuit-of-airmanship-excellence fold as we can. After all, we share the skies with other pilots. Those pilots, if they're not pursuing airmanship excellence, can pose a threat to our safety and to the safety of those we take flying with us. Therefore, it is our duty to recruit as many pilots to the practice of airmanship excellence as we can.

We want to bring aspiring airmen into our airmanship-excellence community because our economic self-interest is served when the cost of our personal flying is reduced by an ever-increasing number of personal flyers. We all learned in Econ 101 that as the customer base for a product or service decreases, the unit cost of the product or service increases. This has been amply demonstrated over the past thirty-plus years. The direct costs associated with personal flying (flight training, aircraft rental, fuel, etc.) have increased by approximately 30% more than the rate of inflation. This effect is even more dramatically demonstrated in the cost of new private aircraft. The annual production rate of new single-engine aircraft has fallen from over 25,000 per year to under 1,000 while the cost of one of these aircraft has increased by approximately 70% more than the rate of inflation.

However, in my opinion, the ultimate motivation for pursuing airmanship excellence is the knowledge that when you're in command of an airplane, you are responsible for your life, the lives of your loved ones, the lives of the passengers you carry on your flights and the people below your flight path that could be hurt or killed if you crash into them. This is a responsibility that cannot be taken lightly.

Almost anyone with reasonable hand-eye coordination, decent reflexes and a good dose of intelligence can learn to drive an airplane. But the personal-flying accident rate over the past three decades has shown that it's not enough to just know how to drive an airplane. You must achieve and maintain a much higher level of airmanship if you want to have the best chance of surviving in a terribly unforgiving and highly dynamic environment.

That's why I believe that if you're going to fly, you should do it right. And Airmanship 2.0 defines "doing it right". In my humble opinion, if you're not prepared to fly right, you shouldn't be flying airplanes by yourself at all. On the other hand, if you're not prepared to invest the time, effort and money it takes to "do it right", you can still enjoy the thrills, insights and other rewards that come from learning to fly and traveling by private aircraft if you're willing to let someone else who is pursuing airmanship excellence do the flying for you. That's "doing it right" too. In the following chapter, you'll learn more about the various methods you can use to pursue airmanship excellence in the right way, but first, let's define "airmanship excellence".

Airmanship Excellence Defined

Before we delve too deeply into the fine points of airmanship excellence and the pursuit thereof, let's define some terms. The goal of this section is to achieve alignment with you on what it is we're actually chasing when we say that we are pursuing airmanship excellence. I discovered during my research for this book that airmanship is not as well defined as one might expect for something that has been around for well over 100 years. One dictionary defines it as *the knowledge and ability needed to control and navigate an aircraft*. That certainly is a concise definition. However, the meanings of knowledge and ability need to be explained in greater detail to precisely define what airmanship is. And this definition leaves out other inherent airmanship attributes like judgment, situational awareness and self-awareness.

Tony Kern, in his book ***Redefining Airmanship***, tells us that most aviators can't provide a comprehensive definition of good airmanship, but they "know it when they see it." Kern goes on to say that *expert flyers are said to have good hands, judgment, common sense and situational awareness*. That helps, but, in my opinion, that isn't precise enough for our purposes. I believe that a concise and comprehensive definition of airmanship for personal flyers can be found in what I refer to as the **Kern Airmanship Model** (see illustration on page 118). Kern developed it while he was Chief of Cockpit Resource Management (CRM) Plans and Programs for the USAF Air Education and Training Command.

The Kern Airmanship Model is made up of ten elements: three bedrock principles, five pillars of knowledge and two capstone outcomes. We'll take a detailed and wide-ranging look at each of these elements later in the book. If you would like to find out more about it before reading any further, you can view the **Kern Airmanship Model In Detail Wings Program Webinar** video briefings on the Center For Airmanship Excellence Website.

If we use the Kern Airmanship Model as a framework for a definition of airmanship, then we can say that *airmanship is the use of sound judgment that is based upon true situational awareness*. Situational awareness is achieved through knowledge of yourself, the airplane you're flying, the team that supports you, the environments within which you are operating and the risks you are facing. These pillars of knowledge are underpinned by proficiency, skill and flight discipline.

The definition of airmanship excellence that I will be relying on for the remainder of this book is: *Airmanship excellence is the possession of airmanship knowledge and skills that exceed those demanded by the type of flying being undertaken.* In other words, airmanship excellence is achieved when the airman's knowledge and skills are sufficient to meet or exceed the demands placed on him at any given time by the highly dynamic flight environment in which he's operating. The consistent adherence to flight discipline, the acquisition of the skills required for the type of flying an airman is doing and demonstrated proficiency in applying those acquired skills are true indicators of airmanship excellence.

Additionally, in my opinion, airmanship excellence cannot be achieved without the ability to identify, assess and mitigate the risks associated with any given flight. And airmanship excellence cannot be realized without the airmanship knowledge and skills needed to foresee and avoid circumstances wherein the personal airmanship capabilities required by the situation may exceed the personal airmanship capabilities possessed by the airman at that time. I also believe that true airmanship excellence cannot be achieved without the support provided by an Airmanship Development Support Organization (ADSO). In other words, I don't believe an individual airman can achieve true airmanship excellence without the backing of a team of aviation professionals who are dedicated to supporting those aviators who are pursuing airmanship excellence. I'll talk a lot more about ADSOs later in this book. If you want more information now, you'll find it in the **Airmanship Development Support Organization** video briefing on the Center For Airmanship Excellence Website.

The below illustration provides another way for us to look at airmanship excellence. I believe the FAA originally produced the chart, but it is now readily available on the Internet. The timeline of a typical flight is shown on the horizontal axis. It covers the timeframe from preflight preparation through taxi, takeoff, cruise, approach-and-landing and taxi in. The curve labeled "Task Requirements" shows the magnitude of the task requirements experienced by the pilot during the various phases of the example flight. The curve labeled "Pilot Capabilities" at the top of the chart shows the level of airmanship capabilities possessed by the pilot in this example. And the area between the task requirements and the pilot capabilities shows the margin of safety this pilot enjoyed on this particular flight.

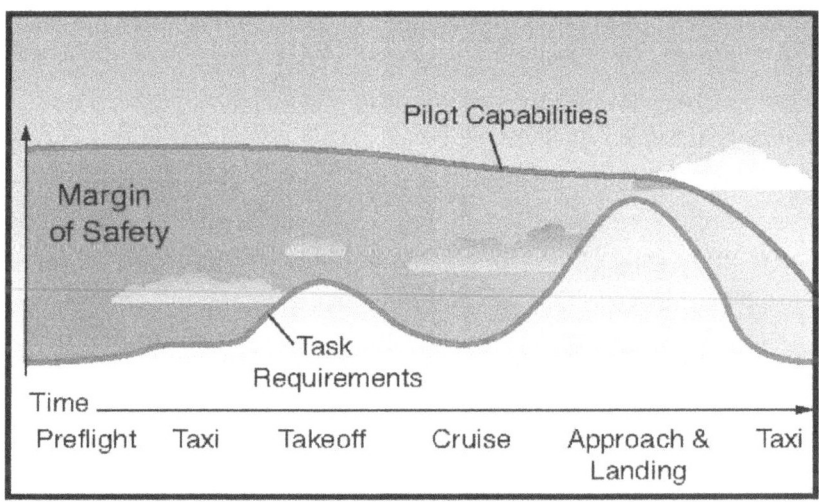

As you will recall, the definition of airmanship excellence I'm using in this book is: *Airmanship excellence is achieved when the airman's knowledge and skills are sufficient to meet or exceed the demands placed on him at any given time by the type of flying being undertaken.* Therefore, we can see in this chart that this particular airman achieved airmanship excellence for this flight since he always had a margin of safety between the demands placed on him and the airmanship knowledge and skills at his command during this flight.

This chart could depict a flight by a personal flyer on a nice day in a simple airplane at a quiet airport, or an airline flight made by a senior captain. In the case of the personal flyer, the task demands placed on him were nowhere near as great as those placed on the airline captain, but our personal flyer still achieved airmanship excellence, according to our definition, because he was prepared to handle all of the demands placed on him and still maintain a margin of safety. The level of airmanship knowledge and skills possessed by the personal flyer may not have been as high as those of the captain, but they were at an "excellent" level when compared to the task loading he faced on this flight.

111

Our personal level of airmanship skills and knowledge will vary over time. Our focus is on achieving and maintaining a level of airmanship knowledge and skills that is appropriate to the kind of flying we're doing. For example, I maintain a level of airmanship knowledge and experience that is appropriate for the kind of flying I'm currently doing—giving flight instruction in single-engine aircraft. If I were to be asked to once again command the left seat of an airliner, I would have to bring my airmanship knowledge and skills needed for that type of flying back up to speed through study and training before I would accept that challenge.

The above should clarify for you what I'm talking about when I use the term "airmanship excellence". You may not as yet be aligned with me on that definition. It is only important that you consider the above definition of airmanship excellence as the basis for the explanation of it that follows.

The History of Airmanship 2.0

In my opinion Airmanship 1.0 began in the time of the ancient Greeks with their mythical tale of **Daedalus and Icarus**. As the story goes, King Minus imprisoned Daedalus and his son Icarus on the island of Crete over a homeland-security issue. Daedalus dreamed up an idea to escape with Icarus from Crete by constructing wings from feathers and wax that he and Icarus could use to fly to safety. He built the wings, and before the two set off he warned Icarus not to fly too low "lest his wings touch the waves and get wet", and not too high "lest the sun melt the wax". But the young Icarus, overwhelmed no doubt by testosterone poisoning, did not heed his father's warning, and flew too close to the sun. And you guessed it, the wax in his wings melted, the feathers departed his wings and he fell into the sea and died.

This is surely a cautionary tale about airmanship. It involves a serious lack of knowledge of self, aircraft, team, environment and risk—the five pillars of knowledge in the Kern Airmanship Model. Unfortunately, it is also the tale of how most personal flyers practice airmanship today. It is my observation that today's personal fliers' overall airmanship knowledge and skills are, for the most part, alarmingly deficient—just like Icarus' were. And of even more concern, today's personal flyers are largely unaware of their airmanship weaknesses.

Icarus was obviously overconfident in his abilities to fly his wings—a trap that many aviators, young and old, fall into. If Icarus had been practicing Airmanship 2.0, he would have been trained in how to recognize and overcome his airmanship deficiencies. Icarus also did not have adequate knowledge of his aircraft's flying qualities and limitations. If he had been practicing Airmanship 2.0, this knowledge would have been baked into his initial and recurrent training. Icarus did not heed his father's advice. He clearly did not know how to use his support team to keep him out of harm's way. Mentoring and shared decision-making are core Airmanship 2.0 principles.

Icarus undoubtedly did not have an acceptable knowledge of the environment he was flying in. After all, he flew too close to the sun and experienced the structural failure of his wings. The practice of Airmanship 2.0, wherein knowledge of environment is a key demonstrated capability, would have forestalled the accident. And Icarus, as so many immature aviators of all ages frequently demonstrate, did not methodically identify, assess and mitigate the risks he faced. If he had been practicing Airmanship 2.0, he would have used a **Flight Risk Assessment Tool** to do that job.

This dichotomy between what Icarus did and what he should have done illustrates the differences between Airmanship 1.0 and Airmanship 2.0. Airmanship 1.0 is "not flying right" and Airmanship 2.0 is "flying right". And the consequences of not flying right are demonstrably less appealing than the aeronautical outcomes experienced by airman who practice Airmanship 2.0.

In the mid-to-late 1400s, Leonardo Da Vinci laid the foundation for Airmanship 1.0 with his twenty-five-year study of aerodynamics and aviation structures. He was also an enthusiastic promoter of personal flight. He is quoted as saying, "For once you have tasted flight you will walk the earth with your eyes turned skywards, for there you have been and there you will long to return." To my mind, this quote demonstrates Da Vinci's intuition about the role "human factors" plays in motivating airman to, or inhibiting them from, making the right decisions. However, this was long before the human race had any actionable evidence that human factors play such an important role in aeronautical decision-making. Therefore, when Da Vinci dreamed about flying, he thought about it in Airmanship 1.0 terms which were more focused on "driving the airplane" than on airmanship excellence.

In the late 1800s and early 1900s, modern aviation pioneers like Otto Lilienthal and Octave Chanute further developed airmanship. However, they were focused on how to build and "drive" an airplane. Once again, this is indicative of Airmanship 1.0 thinking. It was still very early on in the flying game at this point, and the vast majority of airmanship knowledge just hadn't been discovered yet. The early aviation pioneers were severely handicapped by this lack of knowledge. As a result, Lilienthal was killed in the crash of one of his hang gliders.

Although today we have a well-stocked storehouse of airmanship knowledge, the aviators who are still practicing Airmanship 1.0 almost universally suffer from a personal lack of this knowledge because they are unmotivated to put the necessary effort into acquiring and maintaining it. This is the case even though it is far easier to access that knowledge than at any time in the history of aviation. The possession of adequate airmanship knowledge is just as critical to safe flight as it has ever been.

In the late 1890s, the Wright Brothers (Orville and Wilbur) initiated their study of airmanship when they began to codify some of the basic knowledge and skills that would be needed to successfully break the "surely bonds of Earth". They studied, researched and wrote about the knowledge of aircraft, self, team, environment and risks that were needed to fly like a bird. As we all know, they were extremely successful in this endeavor when in 1903 they conducted the first flight of a powered, controlled, heavier-than-air aircraft. However, although they were meticulous in recording their findings and they were certainly moving in the right direction, the type of airmanship they practiced, and the type they taught to their eager students, was rudimentary at best.

Aircraft technology progressed rapidly from the Wright Brother's first powered-and-controlled flight in 1903 through the start of World War I in 1914. The development of Airmanship 1.0 progressed in parallel. Airmen were learning through the school of hard knocks that they had to get better at planning and preparing for their aeronautical adventures. They also witnessed and heard about the results of poor flight discipline and unnecessary risk taking.

I believe that this period, from around 1905 through 1913, was the crucible within which the first Airmanship 1.0 culture was created. Flying at that time was a very, very risky business. Crashes were common and the chances of being killed in one of those mishaps were extremely high. Therefore, the endeavor put out a clarion call to young men (almost exclusively) with a large appetite for risk and an irresistible urge to prove themselves against any challenge. And that is exactly the type of fledgling airman that the new world of aviation attracted.

The composite personality of these "daring young men in their flying machines" consists of a very high degree of self-confidence, more than a touch of arrogance, almost unbounded enthusiasm and an uncanny ability to accomplish the mission even when confronted with insurmountable obstacles. To meet the challenges they faced, these new airmen had to have the ability to master a rather complex piece of equipment that was operated in a highly dynamic environment, and they also had to be able to assimilate knowledge in a broad range of subjects quickly and effectively. Their lives literally depended on it.

These pioneer airmen did a very good job of developing state-of-the-art airmanship knowledge and skills. However, their ability to avoid serious accidents and to utilize their aircraft in a more effective manner suffered from the limited storehouse of airmanship knowledge that existed at the time. So, on the eve of World War I, the Airmanship 1.0 culture of the time was populated by airmen who were intensely dedicated to flying and willing to take on a "devil-may-care" persona in the face of a reality that demonstrated daily that the chances of an airman living to a ripe old age were virtually non-existent.

During World War I (1914-1918), the military services of the combatants advanced the art and science of airmanship at a rapid pace. Pilot-selection processes were established and more-formal pilot training was introduced and practiced on a large scale. The role of pilot-support teams was expanded and the birth of a proper flight-operations organization took place. By the end of the conflict, airmanship had advanced considerably. Following World War I, the Airmanship 1.0 practices that had evolved during the war were transferred from the world of military aviation to the civilian arena. It wasn't long before all pilots—military, airline and personal flyers—were applying to their flying the principles of Airmanship 1.0 developed during World War I.

In fact, towards the end of World War I, the U.S. government established the U.S. Airmail Service (May 1918). Most of the aircraft used in the new service were World War I surplus and most of the pilots who flew those aircraft were veterans of the war to end all wars. Surplus military aircraft also found homes with barnstormers, flight schools and personal flyers. Again, most of the pilots who flew these aircraft were former military pilots. This phenomenon served to spread Airmanship 1.0 principles as practiced by the military in World War I throughout the U.S. pilot population.

By the end of World War I, the Airmanship 1.0 culture that was born before the war had evolved somewhat to include the airman-as-warrior aspect that was popularized during the war. Now, airmen were seen as "daredevils", "dashing" and "knights of the air". In other words, aviating was seen as a high-risk/high-stakes business. And, the support structure for that airmanship culture was rickety at best. Formal aviation regulation was in its infancy, and outside of the military, there were no large-scale organizations to which contemporary civilian airmen could turn to for information and support.

In the years between World War I and World War II (1918-1939) aviation technology advanced rapidly. However, Airmanship 1.0 did not. On the eve of World War II, most personal flyers were still approaching airmanship the way they did when World War I ended. Radio navigation and instrument flying were practiced routinely in the military and airlines, but personal flyers still relied on the basic-airmanship tradition commonly referred to as "IFR—I Follow Roads". Formalized and easily accessible means of increasing an airman's knowledge of himself, his aircraft, his team, the environment he operated in and the risks he faced were still over the horizon.

The airline, military and civilian Airmanship 1.0 cultures that existed as World War II drew nigh were very similar. All three cultures valued the attributes normally associated with "steely eyed" airmen—bravery, perseverance, coolness under pressure, integrity and individualism. None of the cultures put much emphasis on the methodical preparation and thoughtful application of a strict set of airmanship rules, procedures and best practices. Everyone's thinking was more focused on how to "drive" an airplane rather than on how to maximize the safety and efficiency of flight operations. And aviation-human-factors research was still only an embryonic thought in the minds of some aviation pioneers.

World War II (1941-1945 for America) drove the advancement of aviation technology at an astounding rate. Aircraft were designed, tested and built in quantity that could fly significantly higher, faster and farther than their pre-war predecessors. And military airmen learned how to fly them effectively to accomplish their all-important missions. The airmanship knowledge and skills military airmen acquired in World War II, and the ways they developed them, began to lay the foundation for the transition to Airmanship 2.0.

The vast amount of knowledge these World War II airmen assimilated into their daily airmanship far exceeded that required of earlier airmen. They created, learned and developed long-range radio navigation, long-range aerial Celestial navigation, airborne-radar navigation and target detection, crew-coordination techniques, high-quality and highly standardized crew selection and training processes, large-scale aviation-organization-deployment strategies and standardization policies. The most important airmanship-related development to come out of World War II, in my opinion, was the founding of the field of aviation-human-factors research. It was born out of the frustration that the machines of war were becoming so complex and demanding that they were outstripping the abilities of military personnel to safely and effectively manage them. Most of the work during the war was done in aviation systems—training devices, aircrew equipment and instrument-and-cockpit design.

In the early 1980s, I worked with a psychologist in Houston on a project to create a computer-based impairment-testing system. Whenever we had time, I would enjoy listening to his World War II stories. He was a flight surgeon in a U.S. Army Air Corps B-17 wing that flew out of England. He confessed that although he and his colleagues knew they were breaking new ground with their research into aviation-human factors, they did not at that time realize how important that work would be to future aviators.

The airmanship culture that evolved in U.S. military aviation during World War II looked a lot like that of World War I. But added to the "steely eyed" attributes that dominated the culture before the war were qualities like competency, teamwork and a more purposeful approach to airmanship. Following World War II, thousands of military aviators assimilated into the airline, business and personal-flying communities. It wasn't long before the airmanship advances that were made during the war were adopted by military, airline, business and personal flyers alike. Throughout the 1950s, 1960s and 1970s, the pilots in these subcultures approached airmanship in approximately the same way due to this robust cross-fertilization between military, airline, business and personal-flying cultures.

117

Due to the lessons learned by the military in World War II and discoveries in aviation human-factors research during this period, civilian flight training got progressively more organized, standardized and effective throughout the 1950s '60s and '70s. Also airmanship best practices, like the universal use of checklists and standardized traffic patterns, became the norm among personal flyers.

The U.S. government's regulation of aviation began with the passage of the Air Commerce Act in 1926. Many sound, forged-in-blood airmanship-related regulations were enacted between then and the start of World War II. When these rules were complied with, airmen of that era enjoyed an extra margin of safety. From shortly after World War II through the 1970s, the new Federal Aviation Agency, which began life with the Federal Aviation Act of 1958, put scores of new and updated Federal Aviation Regulations (FARs) on the books. Although pilots with "rogue-aviator" tendencies often complained about the expanding reach of these regulations, they in fact made flying much safer if they were followed religiously.

The decades following the end of World War II saw the continuation of military-sponsored aviation-human-factors research. Military research laboratories established during the war were expanded and the Army, Air Force and Navy developed additional labs. Universities also established aviation-human-factors laboratories. Private aviation companies like Boeing established human-factors and ergonomics groups.

By the late 1970s, the airlines were also beginning to look to aviation-human factors research for clues on how to stem the rising tide of jet-airliner accidents. United Airlines was seriously looking at ways to reduce its accident rate in the wake of several high-profile airliner crashes that were attributed to human-factors issues. The "straw that broke the camel's back" for United was the 1978 Portland, Oregon crash of a **Douglas DC-8-61** that ran out of fuel about seven miles short of the runway and crashed into a heavily wooded area. The aircraft ran out of fuel because the crew mishandled the management of an emergency situation.

The National Transportation Safety Board determined that the probable cause of the Portland accident was, "the failure of the captain to monitor properly the aircraft's fuel state and to properly respond to the low fuel state and the crewmember's advisories regarding the fuel state. This resulted in fuel exhaustion to all engines. His inattention resulted from preoccupation with a landing gear malfunction and preparations for a possible landing emergency. Contributing to the accident was the failure of the other two flight crewmembers either to fully comprehend the criticality of the fuel state or to successfully communicate their concern to the captain."

In addition to this United Airlines accident, roughly one dozen other U.S. carriers suffered serious accidents during the 1970s. This prompted the FAA in the late 1970s to mandate Cockpit Resource Management (CRM) training for all U.S. airline pilots. United was first on the list. The FAA ordered United to train all of its pilots in this new approach to airmanship by the end of 1981. It was also in the early 1980s that the military signed on to CRM training as an effective way to lower its aviation accident rate. The airline and military CRM training was based heavily on the aviation-human-factors research that had been going on over the previous four decades.

By the mid-1980s, virtually all U.S. airline and military pilots had been through initial CRM training and subsequently through annual recurrent CRM training. The airlines experienced far fewer serious airmanship-related accidents (about half as many) in the 1980s as they did in the 1970s. Some portion of the reduction in accidents can be attributed to factors like technology and mechanical-reliability improvements, but by this time over eighty percent of all accidents were being attributed to pilot error. This significant reduction in the number of airline accidents took place in the decade after airline deregulation when the number of airline flights increased dramatically.

The National Transportation Safety Board publishes annual airline-accident statistics. The statistic that is generally considered to be the best, but not perfect, yardstick for measuring the rate of aviation accidents is the number of accidents per 100,000 hours flown. In the 1970s, the number of airline accidents per 100,000 hours flown was around 1.4. During the 1980s, it was approximately 0.2 accidents per 100,000 hours flown. This was a reduction of over eighty percent in the accident rate. Most of us senior aviators believe that the lion's share of that reduction was directly attributable to the airline-pilot-culture change that took place during the 1980s.

As I mentioned in Chapter 1, the timing of my airline career put me into the thick of that major pilot-culture change. That change resulted in the acceptance and whole-hearted implementation of human-factors-based CRM principles throughout the U.S. airline and military pilot communities. In many ways, it was a difficult change to make—especially for our more seasoned brethren. It took several years and it cost a lot of money. But by the end of the 1980s, air travelers, airlines, insurance companies, banks and airline pilots all had benefited greatly from the hard-won struggle. You can find a comprehensive explanation of the history of CRM on **FAA TV**.

The new approach to airmanship that came out of this culture change was significantly different from the approach to airmanship that had evolved from the dawn of the air age to that point. It added human factors to the knowledge-and-skills that airmen had to acquire and maintain if they aspired to being an excellent airman. And it embraced the technological changes that were making flying safer and more efficient. This new approach to airmanship is what I call "Airmanship 2.0" and it was already well established on its divergent course away from Airmanship 1.0 by the end of the 1980s.

Over the past three-plus decades, major-airline and military aviators, flight-operations managers, academics, human-factors experts, engineers, inventors and entrepreneurs have continuously refined Airmanship 2.0 to the point where traveling on a major U.S. airline today is by far the safest form of travel ever devised by humankind. However, Airmanship 2.0 has been only partially implemented at most regional airlines so far, and the regional accident rate is about twice that of the major airlines. However, the regional-airline accident rate it is still far below that for personal flying. I believe that the data makes it abundantly clear that the practice of Airmanship 2.0 results in significantly reduced aviation-accident rates to the extent it is comprehensively implemented.

Unfortunately, the personal-flying community is still stuck in the practice of Airmanship 1.0. I believe that it is also true that the practice of Airmanship 1.0 has reached a point where it can no longer further reduce the accident rate for personal flyers. That's certainly what the historical data shows. In 1980, the personal-flying accident rate was 9.9 accidents for every 100,000 hours flown. In 1990 it was 7.9. In 2000 it was 6.6. In 2005 it was 7.2 and it 2010 it was 7.2. The accident rate for any particular year throughout this thirty-plus-year period varied only slightly above and below the average for the period.

From the 1970s to today, the **FAA**, the **Aircraft Owners and Pilots Association**, the **Experimental Aircraft Association** and many other airmanship-oriented organizations have done a great job of packaging and delivering airmanship knowledge. However, as we've seen this hasn't really affected the personal-flying accident rate in any meaningful way, and it appears that the industry is only going to keep doing more of the same for the foreseeable future. Yet it should be clear to all stakeholders that if you want to make personal flying as safe as airline travel, or even significantly reduce the personal-flying accident rate, you have to start doing something different. That something different is clearly Airmanship 2.0. It unequivocally works for the airlines and the military, and I believe that it can work just as well, or maybe even better, for personal flyers.

However that conclusion begs the question: Can Airmanship 2.0 for personal flyers be created and implemented in a manner that it is affordable and accessible for them? I've heard the argument made that Airmanship 2.0 won't work for personal flyers because it will cost too much or the necessary support structure will be too difficult to implement and maintain. Some point out that airline and military pilots are required to practice Airmanship 2.0 by their employers. How are you going to get personal flyers to adopt it if they don't have to and there are cost and effort barriers to practicing it? And airline and military pilots are paid to go to company-paid-for initial and recurrent training. How are you going to motivate personal flyers to complete and pay for all the training that is required by the practice of Airmanship 2.0 if they don't have to?

These are valid questions that I'll deal with in more detail later. In fact, they are the questions I started with when I began looking at solutions to the major problems standing in the way of transitioning personal flyers to Airmanship 2.0. After working on the problem for over four years now, I've come to the conclusion that there are solid answers to these questions and the scores of others that need to be answered to make the practice of Airmanship 2.0 by personal flyers a reality.

I am firmly convinced that the practice of Airmanship 2.0 by the vast majority of personal flyers can be a reality within this decade and that we can and should start making the conversion now. And I also believe that if personal flyers practice Airmanship 2.0 as it is described in this book, the accident rate for those airmen will be even lower than that for airline pilots. I know that these are two rather bold statements with which you may at this point disagree.

121

So, I'd like to encourage you to keep reading to see how I think it can be done. For now, let's focus on the important elements of Airmanship 2.0.

Airmanship 2.0 For Personal Flyers

Airmanship 2.0 for personal flyers is made up of many important essentials. An individual airman can assimilate many of these components on a piecemeal basis. However, even though this may result in an improvement in the airman's airmanship, it will not lead to a genuine pursuit of airmanship excellence or result in the practice of Airmanship 2.0. Airmen who are truly committed to the pursuit of airmanship excellence strive to incorporate every element of Airmanship 2.0 into their airmanship repertoire. So, let's take a look at the elements of Airmanship 2.0 that pertain to personal flying.

I, like many other airmen who have been pursuing airmanship excellence, consider **Gen. Charles "Chuck" Yeager** to be an icon of airmanship excellence. In 1947, then twenty-four-year-old USAF test pilot Yeager broke the sound barrier for the first time in the **Bell X-1**. Just six years earlier (1941), he had enlisted in the U.S. Army Air Forces as a private initially serving as an aircraft mechanic. After little more than a year as an enlisted man, he was accepted into flight training. In March of 1943, he graduated from initial flight training with his wings and a promotion to Flight Officer (the equivalent of today's Warrant Officer). He served as a fighter pilot in Europe where he rose through the ranks to lieutenant and then captain by the end of the war in 1945. By the time he retired in 1975, he was a brigadier general.

Gen. Yeager has been quoted as saying, "If you're going to fly, do it right." This quote was attributed to him early on and he certainly adhered to that philosophy for the rest of his flying career. The idea of "flying right" is an important aspect of Airmanship 2.0. Implicit in the idea is the requirement to know what "flying right" means. The explanation of Airmanship 2.0 that follows clarifies what flying right means to me.

Any current or aspiring airman who undertakes the pursuit of airmanship excellence will first and foremost have to commit himself to the study of aviation human factors. This is a core commitment for the practice of Airmanship 2.0 and "flying right". After all, over seventy percent (some say over eighty percent) of all aviation accidents are attributed to pilot error. Pilots are the type of people who don't like or want to make mistakes. Then what causes us to make these sometimes-fatal errors? When we look into the causes and contributing factors related to accidents attributed to pilot error, in every case the "cause behind the cause" is related to aviation human factors.

I'll go even further to say that in my opinion, one or more of the links in the error chain that leads to every aviation accident today has an aviation-human-factors component. Sure we still have mechanical failures in flight, but they only provide one link in the error chain that is managed by a human. And as we all know, humans make mistakes.

For example, the inflight mechanical failure of the only engine on a single-engine airplane is certainly a critical link in an error chain. However, we also have to look at what caused the engine to fail. More often than not, the failure results from a human error that was made in the course of the engine's manufacture, maintenance and/or operation. FAA-certificated Aviation Maintenance Technicians (AMTs), just like pilots, don't like or want to make mistakes. When we look at what causes them to do so despite intense cultural, regulatory and legal pressures, we always find aviation human factors as the cause of the AMT's error.

And that inflight engine failure doesn't necessarily have to result in an accident. If it's handled properly and at least a little luck is involved, the airman could certainly land the airplane on a runway or in an open field with no damage to the aircraft or passengers. This happy outcome would not require that the airman even file an accident report and the incident wouldn't be classified as an accident. But if this desirable result is to be achieved, the airman at the controls of this airplane will have to fully understand and be in control of numerous aviation human factors.

The airman will have to be at the top of his game. To achieve that, he will have to be unimpaired by fatigue, medications, distractions or stress. He will also have to have a level of airmanship knowledge and skills that meet or exceed the demands placed on him by the engine failure. And he will have to have been prepared for the unexpected engine failure through training, preflight planning and situational awareness.

Some people believe that higher cognitive skills like situational awareness and judgment can't be taught. They believe that you are either born with them or not. However, Airmanship 2.0 as practiced by the airlines and military includes formal initial and recurrent training that has been shown to be effective in teaching the methods and techniques that lead to these desired outcomes. Those training programs have been made possible by aviation-human-factors research. It is primarily through the study of the "Zen" of airmanship—aviation human factors—that one truly achieves airmanship excellence.

Airmanship 2.0 is a new paradigm in airmanship for personal flyers. It is not the way they are flying now. However, as we saw earlier, airline and military pilots have practiced it for over thirty years. Airmanship 2.0 *for personal flyers* is based on the tried-and-true principles of Airmanship 2.0 as practiced by airline and military pilots. Certain elements have been modified for use by personal flyers where necessary. Airmanship 2.0 for personal flyers has been designed to qualify them to operate in today's aviation environment, not the bygone aviation setting Airmanship 1.0 was designed to function in.

This new personal-flying model also provides new ways for airmen to enjoy flying. Many current personal flyers have lost their zest for flying. This is often due to the increasing burdens placed on them by regulations and the necessity to manage a complex undertaking like personal flying. And if their personal-mobility needs are not great enough to generate a large requirement for personal-flying, they are often at a loss for a good reason (in their minds) to go flying. Airmanship 2.0 supports personal flyers in ways that lessens the burdens of making sure all the requisite boxes are checked on the airman's personal list of things he must do if he wants to continue to fly. It also provides multiple motivations for airmen to fly.

For example, the **Airmanship Development Support Organization** (ADSO) that an Airmanship 2.0-practicing airman belongs to will track all of the ancillary requirements the airman must adhere to (i.e., periodic FAA medical examinations, training requirements and aircraft qualifications) and make all of the arrangements that are necessary for the airman to stay in compliance with all the applicable rules and regulations. Also an ADSO professional flight-planning team will prepare a thoroughly researched flight plan for every flight the airman makes, thus reducing the preflight-preparation burden for the airman.

The **Personal Airmanship Development Plan** that every airman who practices Airmanship 2.0 maintains includes elements like **Airmanship Challenges** that motivate the airman to fly more often by providing exciting and airmanship-developing aeronautical challenges. Also, the practice of Airmanship 2.0 requires airmen to fly often enough to maintain the proficiency that is required for safe flying. The social aspects related to membership in an Airmanship 2.0 culture also significantly add to an airman's enjoyment of personal flying. And the rewards and recognition that are an integral part of the culture meaningfully enhance an airman's overall enjoyment of flying.

Airmanship 2.0 also notably increases the value you get for your flying dollar over that derived from Airmanship 1.0. As you will see later in this book, the overall costs of practicing Airmanship 2.0 can be even lower than those incurred while practicing Airmanship 1.0. But even if they are higher in specific cases, the value received in terms of utility, enjoyment and safety are well worth it. Airmanship 2.0 continuously enhances your airmanship skills, knowledge and capabilities. Airmanship 1.0 does not. Airmanship 2.0 is a sure-fire way for you to become a safer, more-proficient aviator.

Airmanship 2.0 includes your formal flight-and-ground training, your informal airmanship training and your Personal Airmanship Development Plan. It also includes the types of aircraft you fly and the way that you fly them. For example, an airman who practices Airmanship 2.0 will always, when he is taking passengers along with him, select an aircraft that has all of the modern safety features that are appropriate to the flight (i.e., whole-aircraft parachute, air bags, glass cockpit, satellite phone). And the airman will fly that aircraft in compliance with the industry-best-practices that have been adopted by his ADSO.

The goal of making personal flying significantly safer can be reached, I believe, through the pursuit of airmanship excellence. As you'll recall, the definition of airmanship excellence we agreed upon (at least for the purposes of this discussion) included: *Airmanship excellence is achieved when the airman's knowledge and skills are sufficient to meet or exceed the demands placed on him at any given time by the highly dynamic flight environment in which he's operating.*

But what are the airmanship knowledge and skills that we're talking about? The specific airmanship knowledge and skills needed by a particular airman who has achieved and is maintaining airmanship excellence are defined by the type of flying the airman is doing. For example, the airmanship knowledge and skills needed by a personal flyer who only flies on nice, sunny days, stays close to his home airport and who flies alone are significantly fewer, and sometimes quite different than, those needed by an aviator who flies night-and-day, in poor weather, in a high-performance airplane carrying passengers.

Fortunately, it is rather easy for an experienced ADSO training team to clearly define what specific airmanship knowledge and skills each airman in the organization needs for the type of personal flying he is doing. These specific airmanship knowledge and skill requirements are then used as the standards the ADSO employs to qualify an airman for the type(s) of personal flying he wants to do. The focused study, training and practice needed for an airman to meet and maintain those standards are incorporated in an airman's Personal Airmanship Development Plan. The Kern Airmanship Model provides the blueprint for creating the plan.

Airmanship 2.0 uses the Kern Airmanship Model as the scaffold upon which airmanship excellence is built. We can see how the Kern Airmanship Model works as a roadmap to achieving airmanship excellence by taking a closer look at each of its elements. But please keep in mind that the following is only an overview of the model. You will find a more-detailed explanation in the ***Kern Airmanship Model In Detail Wings Program Webinar Archive*** on the CFAE Website, and in Kern's book "Redefining Airmanship".

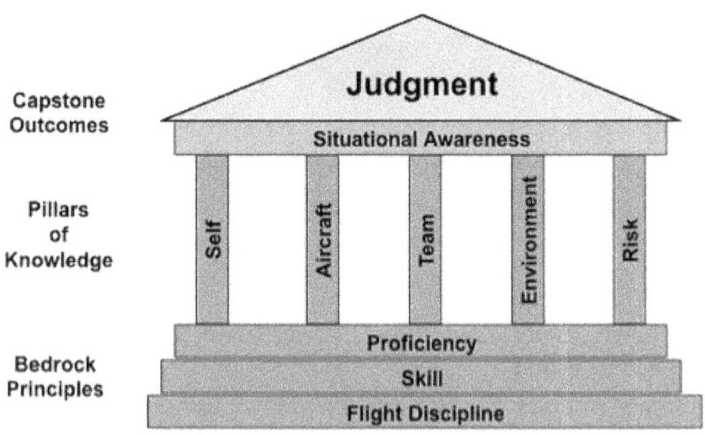

Kern Airmanship Model

Chapter 2: Airmanship 2.0

The Kern Airmanship Model is made up of two Capstone Outcomes (Judgment and Situational Awareness), five Pillars of Knowledge (Self, Aircraft, Team, Environment and Risk) and three Bedrock Principles (Flight Discipline, Skill and Proficiency). The first Bedrock Principle, Flight Discipline, is the ability and willpower to safely fly an aircraft within operational, regulatory, organizational and common sense guidelines. Flight Discipline must be the foundation of an airman's pursuit of airmanship excellence. In my opinion, it is impossible to achieve airmanship excellence without excellent Flight Discipline. If you don't consistently exhibit Flight Discipline, your pursuit of airmanship excellence will be frustrated.

If you have the willpower to begin pilot training, you can muster the willpower to practice Flight Discipline. Your willpower will come from your desire to achieve airmanship excellence. Your ability to practice Flight Discipline will come from your airmanship training and your adherence to personal-flying best practices.

For example, Flight Discipline is exhibited when you unfailingly use a checklist every time you are about to experience a change in phase-of-flight (i.e., before-start checklist, before-takeoff checklist, before-landing checklist), or you have just experienced a change in phase-of-flight (i.e., after-takeoff checklist, cruise checklist, after-landing checklist).

The second Bedrock Principle, Skill, is the ability to do something well. We are not born with, nor do we normally acquire during Airmanship 1.0 training, the airmanship skills we need to be excellent airmen. We must obtain those skills through careful, well-guided and comprehensive training. Airmanship skills give us the true confidence we need to safely and efficiently operate the aircraft we fly.

From an airmanship perspective, confidence comes in three flavors: under confidence, over confidence and true confidence. Under confidence comes into play when an airman decides he can't meet a particular challenge he is facing because he doesn't believe that he has the airmanship skills necessary to meet that challenge when he actually does possess them. Under-confidence will result in poor utilization of the investment an airman has made in his airmanship skills and it may, in some circumstances, lead to poor judgment. Over confidence results from an airman deciding that he can meet an airmanship challenge when his airmanship skills are not in reality sufficient to do so safely. Over-confidence has been a critical link in many accident error chains that have led to a fatal accident. An airman possesses true confidence only when he is certain that his skills are equal to the challenge he is facing. That certitude is valid only if he has recently met that or a similar challenge successfully.

A good example is the challenge of a crosswind landing, an airmanship challenge that can present itself to any airman on any flight. Let's say that our airman wants to land on a runway that has a fifteen-knot **crosswind component**. If the strongest crosswind component the airman has landed in during the past ninety days was five-knots, our airman is operating with over confidence if he attempts a landing with the much more challenging crosswind component. On the other hand, if our airman has in fact landed in a fifteen-knot-or-greater crosswind within the last ninety days, he can have true confidence that he can do it again if his airmanship skills are not compromised by some form of impairment (i.e., stress, fatigue, drugs, alcohol).

True confidence is the valuable reward that is obtained with a high level of airmanship skill. Aviation is terribly unforgiving of inattention, carelessness or neglect. We therefore cannot ignore our airmanship skills. We must know precisely what airmanship skills we need in any given situation and we must know whether or not those skills are sharp enough to meet the challenge we are facing. Unfortunately, the skills human beings acquire deteriorate rapidly if they are not used. We must keep our airmanship skills sharp through recurrent training and practice.

Another problem with airmanship skills is that humans are notoriously bad at evaluating their own skill levels. We have to do our best at assessing our airmanship skills, but we need support from an ADSO to do it properly. This support includes frequent recurrent training, periodic airmanship-evaluation flights with a Mentor Pilot, flight-data collection and interpretation, and feedback from our peers. If we ignore our airmanship skills, as is the norm when practicing Airmanship 1.0, we do so at our own peril. Individual airmanship excellence requires that you seek continuous improvement in your airmanship skills based on a rigorous evaluation of your current strengths and weaknesses.

Flying requires multiple skills that are developed in and out of formal training. In addition to physical flying skills, excellent airmanship requires excellent communications, decision-making, team and self-assessment skills. Airline and military pilots have learned these skills and it is well within the capabilities of personal flyers to do so too.

Airmanship 1.0 flight-training programs are focused on preparing you to reach the government-mandated minimum skill level that is required to pass an FAA checkride. Then the Airmanship 1.0 culture turns you loose, on your own, into a highly dynamic flight environment. You may be able to successfully meet the airmanship challenges you encounter in that environment on the day you pass your checkride, but not long thereafter your skills will have deteriorated to the point where they are no longer up to these minimum-government standards.

Even more alarming is the fact that the Airmanship 1.0 culture requires absolutely no recurrent training. The only requirement is that you receive one hour of ground instruction and one hour of flight instruction from an FAA-certificated Flight Instructor (CFI) every two years. This is certainly not the kind of structured, frequent recurrent training that you need to ensure that your airmanship skills are adequate for the airmanship challenges you will face.

Airmanship 2.0 flight-training programs are focused on fully qualifying you to do the kind of flying you intend to do, and on keeping you qualified through frequent, structured recurrent training. Appropriate levels of FAA certification are milestones on your path to becoming fully qualified, but they are not the end objective of Airmanship 2.0 training. Airmanship 2.0 training is designed to help you to achieve your potential as an aviator. Airmanship 1.0 training is not.

If Skill is one edge of our airmanship sword, Proficiency (the third Bedrock Principle) is the other. Experience and training shape airmanship skills and knowledge. Proficiency is the result of the recent use of the airmanship knowledge and skills that are required to meet the challenges of flight.

Airmanship 1.0 requires pilots to meet certain government-minimum currency-of-flight standards. For example, **Federal Aviation Regulation 61.57** requires that a pilot who intends to carry passengers on a flight must have made three takeoffs and three landings within the previous ninety days in the type of aircraft he intends to fly. If the intended flight is on a day with strong, gusty crosswinds, and if the pilot has not demonstrated his proficiency in those conditions within the past ninety days, the pilot will be legal to fly the flight if he has made his three "bounces" within the previous ninety days. However, in all likelihood his skills will be insufficient to keep everyone on the flight safe. This is a good example of "legal but not safe".

Another example is the fact that under FAA regulations, a pilot does not have to acquire any minimum number of flight hours during a certain period of time to be legal to fly. Yet, the accident statistics clearly show that pilots with less than ten flight hours in the previous thirty days are at far greater risk than those with ten-to-thirty hours. There are many other examples of "legal but not safe".

Airmanship 2.0 is not focused on merely keeping airmen legal to fly. It goes far beyond that to ensure that an airman has the proficiency to meet the airmanship challenges he my face in the type of flying he is doing. Lack of proficiency has been cited as the number one cause of most personal-flying accidents. By ensuring airmen are proficient, the practice of Airmanship 2.0 can significantly reduce the number and severity of the accidents these airmen experience.

The three Bedrock Principles (Flight Discipline, Skill and Proficiency) of the *Kern Airmanship Model* form the foundation for our pursuit of airmanship excellence. However, to reach the heights we want to attain we must build upon this foundation. Kern accomplished this by erecting five Pillars of Knowledge on the three Bedrock Principles: Knowledge of Self, Knowledge of Aircraft, Knowledge of Team, Knowledge of Environment and Knowledge of Risk. Our goal as airmen in pursuit of airmanship excellence is to fill each one of these pillars with the level of knowledge we need to achieve airmanship excellence within the context of the type of flying we are doing.

The first Pillar of Knowledge, "Knowledge of Self", is essential in the pursuit of airmanship excellence given the fact that over seventy percent of all personal-flying accidents are caused by pilot error. Socrates said, "Know thyself." Kern says, "The most critical component of the airmanship model is—yourself." As I mentioned earlier, the study of aviation-human factors and the implementation of the knowledge gained from this research has led to the impressive reduction in airline-and-military flight-operations accidents. *Knowledge of Self* is acquired through the study of aviation-human factors.

And *Knowledge of Self* applied through the practice of Airmanship 2.0 leads to airmanship excellence.

Airmanship 1.0 pilots are more concerned about not looking bad to others than they are about acknowledging the personal human factors that can negatively impact their airmanship. These pilots have a tendency to think that they can rise above the physiological challenges they face even when in fact they can't. This over-confident mindset conflicts with accurate self-knowledge and excellent airmanship.

Airmen who practice Airmanship 2.0 know that while they are responsible for their personal airworthiness, their ADSO is backing them up with tried-and-true Airmanship 2.0 policies, procedures and methodologies that help them to make an accurate assessment of their airmanship capabilities. Our level of airmanship is constrained by our physical and psychological limits. As Kern says, "If we fail to appreciate this fact, we undercut any efforts towards improvement."

"Knowledge of Aircraft", the second Pillar of Knowledge in the *Kern Airmanship Model*, is also critical to the pursuit of airmanship excellence. There is really no such thing as too much knowledge about aircraft in general and the aircraft you are flying in particular when in comes to the pursuit of airmanship excellence. The more you know about how your flying machine is put together and how to operate it, the better airman you are. The basic knowledge that you need to safely and efficiently operate your aircraft can be readily defined, organized and presented to you by a good ADSO training team. Knowledge beyond these basics can best be acquired through a continuous and systematic study of the aircraft you fly.

To acquire this knowledge, you must go beyond mere "book learning" and get your hands a little dirty with the real hardware. A great way for a personal flyer to do this is to spend time as a "mechanic's assistant" when maintenance is being performed on the type of aircraft you fly. You must also learn the "unwritten knowledge" about your aircraft through interactions with other aviators who fly that type of aircraft and by talking with the Aviation Maintenance Technicians (AMTs) who maintain it. You must always be on the lookout for all the ways that can lead you to a more perfect knowledge of the airplane you're flying.

This is true despite the fact that you will probably continually have doubts about your systems and procedural knowledge. I can personally attest to this self-doubt. It seems that although the schools I attend to get qualified on a new airplane are very thorough and professionally delivered, and I pass the final checkride for qualifying to fly the airplane, it usually takes at least a year after that before I feel that I have acquired all of the knowledge I need to operate the airplane efficiently. And I find that I continue to discover new details about that aircraft for as long as I fly it. This is obviously because my standards are much higher than the minimum-government standards I had to meet in the checkride. So, I have to employ a continuous and systematic approach to acquiring this knowledge. But I don't stop there. The pursuit of airmanship excellence requires that I continue to learn more about the airplane as long as I'm flying it.

I believe this self-doubt about *Knowledge of Aircraft* is a good thing. For me, it's not about the things I have to know to operate the airplane safely and efficiently, it's about fearing that I may not know enough about my airplane to extricate myself from an emergency situation that is the result of a mechanical failure. This fear fuels my quest for greater understanding of the airplane I'm flying, and that's a good thing.

The third Pillar of Knowledge in the *Kern Airmanship Model* is "Knowledge of Team". This of course implies that airmen who are practicing Airmanship 2.0 do so with the support and assistance of other people. This is a key distinction between Airmanship 1.0 and Airmanship 2.0. Airmanship 1.0 personal flyers usually operate without the support of an organized team. They may have a team of sorts that is comprised of an AMT, a flight instructor and a ground-service crew. But it is usually ad hoc and not coordinated in a way that can enhance the pilot's safety and airmanship.

The airman who is in charge of a particular flight is commonly referred to as the "Pilot In Command" (PIC). This concept is codified in both the Federal Aviation Regulations (FARs) and the regulations of the International Civil Aviation Organization (ICAO). The FARs tell us that the *Pilot In Command*: "Has final authority and responsibility for the operation and safety of the flight; has been designated as pilot in command before or during the flight; and holds the appropriate category, class, and type rating, if appropriate, for the conduct of the flight." ICAO defines *Pilot In Command* as "The pilot responsible for the operation and safety of the aircraft during flight time."

A reasonable interpretation of these regulations leads one to believe that the *Pilot In Command* is the leader of his flight-support team. United Airlines construed the regulations this way in the early 1980s when it named its initial cockpit-resources-management program "Command/Leadership/ Resource Management". A large part of the training was focused on teamwork and the roles of the leader and followers on the airline-cockpit team.

Of course, effective leadership is the cornerstone of excellent teamwork. Leadership skills are not something we are born with. We learn them informally and through training. Airmanship leadership requires a specialized subset of leadership skills that are learned through Airmanship 2.0 training programs, flight instructors and Mentor Pilots. Airmen who practice Airmanship 1.0 rarely, if ever, receive any formal airmanship-related leadership training.

An ADSO institutes norms of interaction and teamwork for the airmanship-support teams that form within its jurisdiction. The ADSO also establishes standards of safety and compliance that the teams all adhere to. And the ADSO creates an environment that is conducive to excellent teamwork. The Airmanship 2.0 culture of an ADSO must also promote confidence and trustworthiness. Excellent teamwork improves performance in three ways: it adds to the team's pool of knowledge and expertise and makes them available to confront various situations, it facilitates an amalgamation of ideas and skills to make new options available to the group and it enables synergy among the team's members. You can look upon excellent teamwork by an airman's team as his insurance policy.

The fourth Pillar of Knowledge in the *Kern Airmanship Model* is "Knowledge of Environment". Mastery of the knowledge related to an aviator's environment can be the most difficult to achieve since it is complex, dynamic and wide-ranging. The operational environment affects our airmanship safety, efficiency and effectiveness. The environment that an aviator operates in can be subdivided into three types: physical environment, regulatory environment and organizational environment.

The physical environment is the arena within which we practice airmanship. It impacts every aspect of the flights we fly. The weather we confront, the terrain we fly over and the airports we operate from/to are examples of the physical environment we operate in. Some factors in the physical environment remain comparatively constant (i.e., the physical layout of airports, the terrain we fly over). Others, like the weather, are in a constant state of change. Our knowledge of the physical environment we operate in must be complete enough to provide us with the information we need to fly safely and to afford a margin of safety. After over fifty years as an aviator, I still learn new things virtually every day about the physical environment I fly in. In fact, this is one of the aspects of Airmanship 2.0 that I enjoy the most.

The regulatory environment airmen operate in is multifaceted, complex and critical to our survival. It includes international regulations enacted by ICAO, Federal Aviation Regulations, local-airport rules and the procedures mandated by the ADSO that you are a member of. These regulations have been put on the books in response to errors made by airmen and/or the flight organizations they fly within. Airmen ignore them at their own peril. However, many pilots feel that flying has become overregulated. Others disagree citing the need for more-stringent rules. The opinion of individual aviators is really not important because we are all bound by the regulatory environment in which we fly, and other pilots base their actions on the expectation of our compliance.

The character of the organizational environment we operate in is a reflection of the nuts-and-bolts of the ADSO we call home. A good ADSO embraces the concepts and methods from cultural anthropology and business along with proven airmanship practices. An Airmanship 2.0 organizational environment employs the principles of shared decision-making, advocacy and assertiveness. And, it includes a safety culture with a safety-management system. It is our responsibility to learn everything we can about our organizational environment and to be at home in it.

The fifth Pillar of Knowledge in the *Kern Airmanship Model* is "Knowledge of Risk". Unfortunately, there is a fine line between taking a calculated risk and doing something really dumb. In the late 1800s, Count Helmuth Von Moltke, the then Prussian Army Chief of Staff said, "First recon, then risk." Kern says, "...you must understand the other elements of airmanship to understand, assess and manage risk, and you must manage risk to grow in airmanship."

Chapter 2: Airmanship 2.0

Kern defines risk as "the probability and severity of a loss linked to a hazard." The identification of hazards to flight is the first step in the process of mitigating risk. Once a hazard is recognized, the probability of encountering that hazard must be estimated and the gravity of the results of that encounter must be assessed. This process provides us with a close approximation of the risk posed by the hazard. The risk is then managed by avoiding it, reducing it, spreading it, controlling it and/or transferring it. Even with the diligent practice of Airmanship 2.0, flying will never be entirely risk-free. Our ability as airmen to comprehend and examine various risk factors, and then manage them using skill and good judgment will always be one of the standards of airmanship excellence.

To complete the structure of his airmanship model, Kern placed two "Capstone Outcomes" atop the five *Pillars of Knowledge*. The first Capstone Outcome, "Situational Awareness", is the state of awareness possessed by an aviator at any given time that seamlessly integrates what has happened with what is now happening and what might happen in the future. Situational Awareness produces observations and conclusions that are then used in the judgment process to make timely and accurate decisions.

However, as is common to almost every process, if the information going into it is unrealistic or inaccurate, the information produced by the process will be likewise compromised (GIGO: garbage in, garbage out). The *Kern Airmanship Model* makes it easy to see that the information going into the situational-awareness process is dependent upon the quantity and quality of the information contained in an airman's *Five Pillars of Knowledge*. And the proper use of that information depends upon the airman's ability to deploy *Flight Discipline*, *Skill* and *Proficiency*.

Case studies and accident analyses from all areas of aviation tell us that Situational Awareness is essential. In an analysis of all USAF mishaps from 1989 through 1995, lost Situational Awareness was the single largest contributing factor. Another USAF study of air-to-air combat kills estimated that ninety percent of fighter pilots who were shot down never even saw their attacker. Now, that's obviously poor situational awareness.

The following quotes from Kern's "Redefining Airmanship" sum up the importance of maintaining Situational Awareness:

135

> "Situational Awareness is the filter through which every aviator's action must be viewed."
>
> "Situational Awareness feeds aviator decision making and judgment."
>
> "Situational Awareness cannot be viewed in isolation."
>
> "All foundations and pillars of airmanship can support or detract from Situational Awareness."
>
> "Situational Awareness can and should be improved by individual effort."
>
> "Situational Awareness is _not_ an innate quality, but rather a higher-echelon piece of the airmanship whole."
>
> "We can improve Situational Awareness to whatever degree we have the will and discipline to achieve."

Sitting atop *Situational Awareness* in the *Kern Airmanship Model* is the ultimate *Capstone Outcome* of "Judgment". The dictionary defines judgment as "The process of forming an opinion or evaluation by discerning or comparing." Kern says it is "The process of comparing and evaluating courses of action—as defined by the pilot and the extended flight team."

Judgment cannot be considered in isolation. It is the product of the rest of the *Kern Airmanship Model*. In fact, the model must be viewed holistically. The elements of the model must be integrated and they must operate accurately and effectively on many different levels to result in good judgment.

We can see the requirement for good judgment in everything associated with airmanship. Poor judgment has had a significant impact on all areas of personal flying. This impact can be seen in almost every accident study conducted in recent years. In one five-year study of personal flying accidents, decisional errors accounted for more than fifty percent of fatal errors. Judgment is the result of a complex aviation-human-factors calculation. Understanding and cultivating Judgment requires a solid foundation of airmanship knowledge and a process for study and improvement.

Judgment sits at the apex of the *Kern Airmanship Model*. All of the other elements of the model (Flight Discipline, Skill, the five Pillars of Knowledge and Situational Awareness) support Judgment. Through the practice of Airmanship 2.0, we can ensure that the judgments we make are all good judgments. Good Judgment in turn supports the major goals of airmanship excellence: safety, effectiveness and efficiency.

The following quotes from Kern's "Redefining Airmanship" highlight the importance of good judgment:

> *"The foundations and pillars of knowledge of airmanship form the basis for good judgment and decision making."*
>
> *"Judgment is not a magical quality."*
>
> *"It can be taught and learned."*
>
> *"The improvement process demands disciplined attention and constant self-critique."*
>
> *"In the final analysis, our judgment is as good as we want to make it."*

If the *Kern Airmanship Model* is a road map we can use in our pursuit of airmanship excellence, then the directions that go with it are: Faithfully follow the Bedrock Principles of Flight Discipline, Skill and Proficiency. Continuously strive for a more perfect knowledge of Self, Aircraft, Team, Environment and Risk. If you follow these directions, you will come to a state of enhanced situational awareness that will lead you to your destination — excellent airmanship judgment. And you will be pleased to find that you can apply a very similar approach to achieving excellent judgment in other areas of your life.

Since the *Kern Airmanship Model* provides us with such a good blueprint for understanding how we can achieve airmanship excellence, I decided to borrow its structure to create a similar model for Airmanship 2.0 (see page 138). The Airmanship 2.0 Model consists of three Bedrock Principles (Commitment, Integrity and Shared Experience). Upon the three Bedrock Principles, five Pillars of Support (Airmanship 2.0 Culture, Safety Management System, Airmanship Training, Modern Aircraft and Dedicated Airmen) are erected. The Capstone Outcomes of Safety and Enjoyment sit atop the five Pillars of Support.

"Commitment" is the first Bedrock Principle of the *Airmanship 2.0 Model*. If you are going to achieve airmanship excellence, you must first make a solid commitment to its pursuit. Your quest is not a short one. You will be on it for the rest of your flying life. The commitment can't be weak or half-hearted. Without this serious commitment of your focus, time, effort and

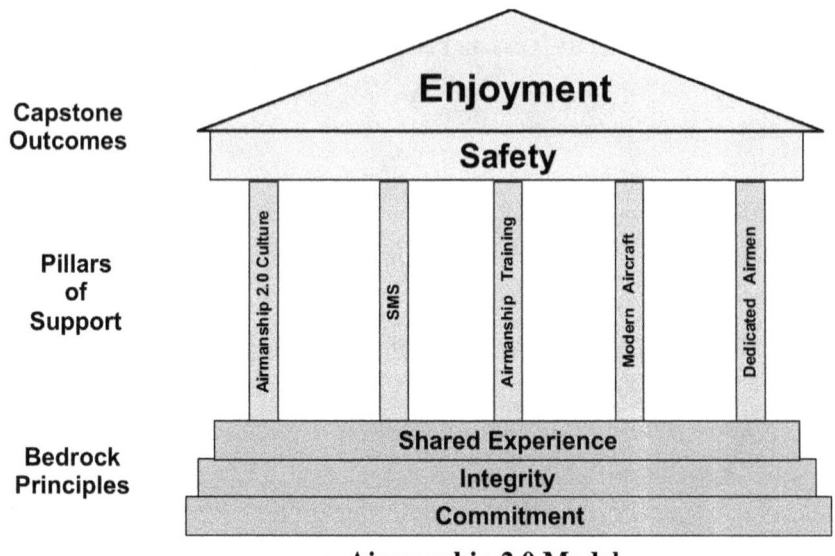

Airmanship 2.0 Model

money, you have little-to-no chance of achieving and maintaining airmanship excellence.

The second Bedrock Principle, "Integrity", must be a central part of your pursuit of airmanship excellence. In the context of Airmanship 2.0, integrity means being honest and open and adhering consistently to the principles of Airmanship 2.0. It also means doing the right thing even when no one else will know whether or not you did it. Integrity means you can be relied upon to never distort information you are sharing with your team members and to not withhold information that may be of use to the team.

"Shared Experience" is the third Bedrock Principle. You cannot fully practice Airmanship 2.0 in isolation. You need to share your experiences with your team and peers, and you need to have them share their experiences with you. This Bedrock Principle can't be eliminated from your pursuit of airmanship excellence. You can't achieve airmanship excellence all by yourself.

The first Pillar of Support, "Airmanship 2.0 Culture" is critical to a successful pursuit of airmanship excellence. The airmanship culture you belong to can either assist or hinder your pursuit of airmanship excellence. An Airmanship 1.0 culture will not support your pursuit of airmanship excellence.

"Safety Management System", the second Pillar of Support, is needed to not only keep you safe when you fly, but also to help you to continuously improve your airmanship. The Safety Management System (SMS) used by the ADSO you belong to must be a formal system. Robust data-collection, hazard-identification, risk-mitigation and safety-communications techniques, policies and procedures must be included in the SMS.

The third Pillar of Support, "Airmanship 2.0 Training", differs considerably from Airmanship 1.0 training. It incorporates leading-edge training systems, processes, procedures and tools to provide airmanship training that facilitates the pursuit of airmanship excellence. This training must be ongoing and comprehensive.

"Modern Aircraft" is the fourth Pillar of Support. Modern aircraft have the improved safety features that make personal flying much safer and more efficient. This doesn't mean that you can't fly older aircraft, only that you need to carefully evaluate the safety risks these aircraft pose (i.e., no whole-aircraft-parachute system, old wiring, fatigued airframes and engines, lack of modern avionics, sub-par crashworthiness and no air bags).

The fifth Pillar of Support, "Dedicated Airmen", is important to your pursuit of airmanship excellence because, once again, you can't pursue it alone. You need the positive and negative reinforcement that only fellow airmen who are also committed to the pursuit of airmanship excellence can provide. These airmen must be dedicated to following the principles of Airmanship 2.0 and to freely sharing their experiences, knowledge and ideas with everyone in their ADSO.

The two Capstone Outcomes I have placed atop the five Pillars of Support in our *Airmanship 2.0 Model* are "Safety" and "Enjoyment". These outcomes are what Airmanship 2.0 is designed to deliver. Airmanship 2.0 has proven that it can produce significant safety improvements for airmen who practice it (i.e., airlines and military). If you and your passengers can feel confident that your flying is as safe as possible, you and they will enjoy the experience immensely.

I believe that we can also use the *Airmanship 2.0 Model* as a road map for our pursuit of airmanship excellence. The directions for following that map are: First, make a sincere commitment to the pursuit of airmanship excellence. Then practice Airmanship 2.0 with unwavering integrity. Share your airmanship experiences with your peers and support team. Always belong to an

Airmanship 2.0 culture. Fully participate in an effective Safety Management System. Acquire and maintain your airmanship knowledge and skills through Airmanship 2.0 training programs and a Personal Airmanship Development Plan. Whenever possible and/or appropriate, fly modern aircraft. Dedicate yourself to following the principles of Airmanship 2.0 and to freely sharing your experiences, knowledge and ideas with everyone in your Airmanship 2.0 culture. If you follow these directions, you will be assured that your flying is as safe as it can be and you will arrive at a true enjoyment of your flying experiences.

Airmanship 2.0 Culture

It is said that culture is all around us. But I'm not talking about the kind of culture we find in fine the arts, music and the theater. I'm talking about the kind of culture that comes from the beliefs, values, assumptions, behaviors and artifacts that are shared by the individuals in a culture. The kind of culture I'm talking about can also be defined as: *The attitudes and behaviors used by a group of people who interact with each other on a regular basis.*

Let's peel the onion back one more layer and define attitudes and behaviors in the context of the Airmanship 2.0 Culture. Attitude is considered to be a settled way of thinking or feeling about someone or something. An attitude will typically be reflected in a person's behavior. We also think of attitude as a truculent or uncooperative behavior. For example, there are five dangerous attitudes a pilot can have: resignation, anti-authority, impulsivity, invulnerability and macho. These attitudes are usually expressed in a resentful or antagonistic manner. The Airmanship 2.0 Culture is designed to identify and discourage these attitudes in a way that supports the pursuit of airmanship excellence. On the positive side, attitudes like teamwork and self-confidence are encouraged within an Airmanship 2.0 Culture. A behavior is defined within the context of the Airmanship 2.0 Culture as: *the way in which one acts or conducts oneself, especially toward others.* From an airmanship-excellence perspective, we define "behaviors" as: *the way in which an airman acts in response to a particular situation or stimulus.*

I'll further define the kind of culture I'm talking about by pointing out that it is an aviation-safety culture that adheres to the principles of Airmanship 2.0. This is an aviation culture with a laser-like focus on continuous airmanship development, safety, efficiency and enjoyment. An Airmanship 2.0 culture is based on the beliefs, values, assumptions, behaviors and artifacts that have been proven to be effective in airline- and military-pilot cultures.

We all operate within multiple cultures. These include family, church, work, professional and national cultures. Family and national cultures are difficult to change because they surround us from the day we are born and they generally have long histories and a powerful inertia. The cultures we encounter at church and work are also resistant to change. The bigger cultures are, the more unwilling they are to modify business as usual.

Professional cultures can be changed if strong incentives drive the transformation. A very good example of this is the radical change in the United Airlines pilot culture following the introduction of **Command/Leadership/Resource Management** training in the early 1980s. This conversion from Airmanship 1.0 to Airmanship 2.0 took place because the FAA mandated it. The mandate provided a strong incentive to change the way airline pilots approach airmanship.

The multiple cultures within which we operate influence us in the cockpit because each of them has an impact on the critical behaviors that we develop as airmen. From an airmanship-development perspective, these cultures all have strengths and weaknesses. Their strengths can assist our pursuit of airmanship excellence and their weaknesses can inhibit it. They also shape our attitudes about risk taking, stress and personal responsibility. Our attitudes about using standard operating procedures, regulatory compliance and open communications are also heavily influenced by these cultures. They also influence how we value and use automation in our flying.

An Airmanship 2.0 Culture can be viewed as both a professional culture and an organizational culture. Even though personal flyers are not professionals in that they are not being paid to fly, their approach to airmanship must be of a professional nature. This professional approach is characterized by the terms "businesslike", "very competent" and "doing something habitually".

From the organizational-culture perspective, an Airmanship 2.0 Culture is a prominent feature of the Airmanship Development Support Organization (ADSO) that supports airmen who practice Airmanship 2.0. This culture includes airmanship-development leadership, a **Safety Management System** and a focus on safety. An Airmanship 2.0 Culture also consists of values, behaviors and artifacts that promote airmanship excellence.

The professional-pilot approach to an Airmanship 2.0 Culture is conducive to airmanship excellence because professional airmen take pride in being an exceptional pilot. This is because airmen who practice Airmanship 2.0 are strongly motivated to "fly right". This stimulus helps an ADSO with an Airmanship 2.0 Culture to work towards airline-like safety and operational efficiency. However, most pilot cultures include a strong negative component that is embodied in a near-universal sense of personal invulnerability. An Airmanship 2.0 Culture includes methods to identify and correct this dangerous attitude.

The majority of pilots who are members of an Airmanship 1.0 culture feel that their decision-making is as good in an emergency situation as it is in normal operations. They also believe that their personal performance is not adversely affected by their personal problems and that they don't commit more errors in situations of high stress. Research shows that these assumptions are simply not true. This misplaced sense of personal invulnerability can result in a failure to utilize best practices as countermeasures against error. An Airmanship 2.0 Culture turns these assumptions on their heads and designs its training-and-education programs to ensure that the airmen in the organization don't fall into this false thinking.

An ADSO (the organization) provides the framework within which national, professional and organizational cultures operate. Its policies and procedures are chief determinants of the behavior exhibited by the members of that culture. The greatest leverage can be exerted to create and nourish a safety culture at the organizational level.

The robust and demonstrated commitment of all stakeholders in an ADSO is needed to establish and maintain an Airmanship 2.0 Culture. Policies that encourage open communication and action (instead of denial) as a reaction to problems and risks is also necessary. Before any organization can build a safety culture, all participants need to decide if they are prepared to take the actions that are required to achieve airmanship excellence, including those that are costly.

The basic conditions that must be met to have a reasonable chance of creating an Airmanship 2.0 Culture include trust, a non-punitive policy towards error acknowledgement and a commitment to action to lessen error-inducing conditions. The ADSO must collect diagnostic data that reveal the kind of threats faced by the organization and the types of errors that occur within it. The ADSO must provide education and training in threat-recognition and error-avoidance-and-management strategies for the airmen in the organization. Training in evaluating and reinforcing threat recognition and error management for instructors and evaluators must also be incorporated in an Airmanship 2.0 Culture.

If safety efforts are to succeed, flight organizations must have a full understanding of the cultural influences on their operations. The Airmanship 2.0 Culture recognizes the importance of culture in its support of its members in their pursuit of airmanship excellence. Culture also serves to bind us together as members of a group and to provide hints and signals as to how to behave in ordinary and new situations. The Airmanship 2.0 Culture binds its members together through shared experiences wherein the leaders of the culture model standardized operational and cultural behavior.

Airmanship 2.0 Culture also explains "the way we do things around here". It codifies what goes and what doesn't. The behaviors that are exhibited in an Airmanship 2.0 Culture reflect certain assumptions about people and how they think and act. It also reflects the values and beliefs that are shared by its members — whether or not they have been articulated.

The values and beliefs of the Airmanship 2.0 Culture are reinforced by artifacts (i.e., icons, stories, heroes, rites and rituals) that remind people of what the culture stands for. The explanation of how things are done in an Airmanship 2.0 Culture is backed up by efforts to measure behaviors and take corrective action when the behaviors are unacceptable to other members of the culture. These assumptions, values, beliefs, behaviors, artifacts, measurements and actions determine how things get done in an Airmanship 2.0 Culture.

Culture can also be thought of as a powerful element of organizational strategy. If it is designed and implemented properly, it can yield impressive returns on investment. Culture is the personalizing force in what would otherwise be an uninteresting and unsatisfying organizational experience. It can nurture trust, create expectations, ease intra-organization communications and reduce uncertainty in relationships between members of the culture. In doing so, it provides better outcomes. The Airmanship 2.0 Culture is an important element in the strategy to reach the goals of a significantly improved personal-flying safety record and of a rapidly expanding number of aviators who practice Airmanship 2.0.

The collective story the members of the group tell each other explains the culture of the group. This story drives the thinking that drives the behavior that is exhibited by the members of the culture. The story includes the values, beliefs, shared assumptions, behaviors and artifacts that describe the vision of the culture that its members share.

The underlying values that a culture shares are integrated into the thinking of its members from the time they join the culture until they leave it. The values shared by the members of an Airmanship 2.0 Culture include: honesty, openness, clarity, alignment, process, integrity, respect, trust, accountability, responsibility and airmanship excellence. Let's take a closer look at each one of the shared values.

Honesty is typically thought of as the quality of being free of deceit and untruthfulness. It is sometimes defined as being sincere. This value is included in the Airmanship 2.0 Culture because every interaction between the members of a the culture must be free of any untrue or misleading information. To do otherwise is to court disaster. Total honesty, applied consistently and continuously, is essential to the personal and cultural survival of the members of the Airmanship 2.0 Culture. Another important reason to include honesty in the shared values of the Airmanship 2.0 Culture is that trust cannot exist when dishonesty is allowed to creep into a culture.

Openness is usually described as being accessible and easily read. When we say that someone is "open", we mean that we trust that he isn't harboring a hidden agenda that may be in conflict with the agenda we think we are interacting with him on. The Airmanship 2.0 Culture requires us to continually improve our openness and never withhold an important piece of information just to serve a personal agenda.

Clarity is the quality of being clear about the information we are using to make critical decisions. This value is important to the Airmanship 2.0 Culture because lack of clarity can lead to bad outcomes. Clarity can many times be difficult for us to achieve in our own thinking, but it becomes exponentially more challenging to achieve among two or more people. As a shared value of the Airmanship 2.0 Culture, clarity is sought at all levels of decision making through the use of methodologies, policies and procedures that make clarity much easier to achieve.

Alignment is a position of agreement or alliance. It is an Airmanship 2.0 Culture shared value because it is critical to good judgment and teamwork. If the members of your flight-support team are thinking in different directions on a critical issue, the thinking of all the members of the team must at some point enter a state of alignment if things are to proceed safely and efficiently to the shared goal of the team. Many aviation accidents have been attributed to a misalignment among crewmembers at a critical point in the accident chain.

Process is another important shared value of the Airmanship 2.0 Culture because it is very difficult to interact with others if we don't know what the rules of engagement are. For example, if you're waiting for me to arrive at our planned 1000 meeting and it's now 1010, your anxiety level begins to rise and you start to think about what you should do—abandon the meeting, wait longer or call me. On the other hand, if we have an agreed-upon rule that I'll call you if I'm going to be more than fifteen minutes late, the uncertainty about what to do is removed. In an Airmanship 2.0 Culture, Process is manifested in operating procedures, checklists, training programs and many other artifacts of the culture.

Integrity is considered to be "internal consistency". What you see is what you get. Integrity is an essential value of the Airmanship 2.0 Culture because unanticipated behavior can in itself be dangerous if it is exhibited at a critical time. Lack of integrity also undermines honesty, openness and trust.

Respect is a feeling of deep admiration for someone or something that is elicited by their abilities, qualities or achievements. Since every member of an Airmanship 2.0 Culture is continuously improving and demonstrating the behaviors that are needed for a successful pursuit of airmanship excellence, the prerequisites for respect are present in the culture. It is easy for Airmanship 2.0 Culture airmen to respect their peers. Disrespect that has been generated by outside influences must not be allowed to insinuate itself into the culture.

Trust is a belief in the reliability, truth, ability or strength of someone or something. For the belief to be justified, a track record must be established between the individuals who are in a state of trust with one another. Within the Airmanship 2.0 Culture, this history is built up through frequent cultural interactions and standard operating procedures. As has been pointed out many times down through the ages, trust if fragile. The Airmanship 2.0 Culture shared values of honesty, openness, clarity and integrity are essential to creating a culture of trust.

Accountability means that every member of an Airmanship 2.0 Culture is answerable for his decisions, actions and behaviors. The fabric of an Airmanship 2.0 culture is made stronger by weaving the thread of accountability through it. The ability to explain one's decisions, actions and behaviors to others in an Airmanship 2.0 Culture is an airmanship skill that is highly valued by the members of the culture.

Responsibility for the safety of a flight is assigned by the FAA to the Pilot In Command of the flight. This is the airmanship equivalent of "the buck stops here". In an Airmanship 2.0 Culture, every member of the culture readily accepts personal responsibility for making sure that the culture reaches its shared goals of flight safety, operational efficiency and expansion of the personal-flying community.

Airmanship Excellence is the central shared value of an Airmanship 2.0 Culture. It is the value that binds us all together and provides a clear reason for the culture's existence. It is what motivates the members of the culture and it defines the culture's purpose.

The underlying beliefs that a culture shares are reflected in the accepted behaviors exhibited by the members of the culture. The beliefs the members of a culture share are taught to the members when they join the culture, and they are continuously reinforced. Effective communication of a culture's shared beliefs seeks to answer two questions for the members of the culture—what's in it for the organization and what's in it for me (the member)?

A belief is the acceptance that a statement is true or that something exists. It is also a firmly held opinion or conviction. Holding a belief demonstrates trust, faith and confidence in the culture. Basic cultural beliefs include the values and principles that guide both strategic decisions and daily behavior. The basic beliefs of quality, people-first, ethics and growth serve as a strong foundation for an Airmanship 2.0 Culture and they will continue to be the basis for future strategy, plans, and accomplishments.

Enjoyment. Members of an Airmanship 2.0 Culture believe that the best way to attract new people to the personal-flying community is to make traveling by private aircraft more enjoyable than other business-and-personal-mobility solutions. They also believe that personal flyers must truly enjoy the personal-flying experience if they are going to stay with it for the long term. Enjoyment of personal flying is a very important component of the value proposition that attracts people to it.

Mutual Respect. Included in the Airmanship 2.0 Culture is the basic underlying belief in respect for the dignity and rights of each person in the culture. This applies to the airmen in the culture as well as the support team. Mutual respect will result in the best customer service to the members of the culture. It will motivate the flight-support team to pursue all tasks with the objective of accomplishing them in a superior way.

Safety. Members of an Airmanship 2.0 culture also believe that the pursuit of airmanship excellence is the best way to keep themselves and their passengers safe. Every member of an Airmanship 2.0 culture must believe that a focus on safety will enhance everyone's enjoyment of the personal-flying experience. Policies and procedures that put safety first must be baked into the culture.

Value. Members of an Airmanship 2.0 Culture must believe that it is the best way to extract maximum value from their personal-flying dollar. No matter how you cut it, personal flying is relatively expensive. However, the actual cost of personal flying is not as important as the value it delivers.

Airmanship Outreach. Members of an Airmanship 2.0 Culture also believe that it will expeditiously expand the pilot population. They realize that increasing the size of the personal-flying customer base will reduce the cost of personal flying. And they believe that networking to potential personal flyers is the best way to bring more people into the personal-flying community.

An assumption is a thing that is accepted as true or as certain to happen, without proof. Shared assumptions provide the basis from which other elements of a culture are derived. A culture can be described as a pattern of shared basic assumptions that was learned by the members of the culture. The members of a culture will accept this pattern of assumptions because it has worked well enough to be considered valid. The underlying assumptions of a culture must be taught to new members of the culture as the proper way to perceive, think and feel in relation to problems they may face. These shared assumptions must be effectively communicated to everyone in the culture.

The members of an Airmanship 2.0 Culture share many basic assumptions. Among them are the basic assumptions that:

Flying is unforgiving. Personal flying is a terribly unforgiving endeavor and it needs to be approached as such.

Airmanship 2.0 leads to airmanship excellence. Practicing Airmanship 2.0 is the best way for an airman to pursue airmanship excellence.

If we demonstrate it, they will come. The continuous demonstration of the efficacy of Airmanship 2.0 will result in its adoption by many other aviators.

Teamwork is insurance. The practice of teamwork within the context of an Airmanship Development Support Organization (ADSO) increases safety and efficiency.

Behaviors. The behaviors exhibited by the members of a culture are driven by the shared values, beliefs and assumptions that are held by them. A properly designed ADSO will make it easy and enjoyable for its members to consistently demonstrate these behaviors. The behaviors that an Airmanship 2.0 Culture is designed to elicit are:

Airmanship Development. Every personal flyer in the Airmanship 2.0 Culture has a Personal Airmanship Development Plan (PADP) that he/she consistently adheres to.

Airmanship Best Practices. Every member of the Airmanship 2.0 Culture applies to his/her personal flying the airmanship practices that the culture designates as best practices.

Regulatory compliance. All members of the Airmanship 2.0 Culture voluntarily comply with all applicable FAA, ADSO and airport rules and regulations.

Teamwork. Every member of the Airmanship 2.0 Culture practices cooperation and collaboration within the context of a flight-support team.

Airmanship Outreach. Networking and referrals are a common practice among the members of the Airmanship 2.0 Culture.

Honesty. The members of the Airmanship 2.0 Culture never lie to one another.

Openness. Everyone within the Airmanship 2.0 Culture freely volunteers information that may be needed by other members of the culture.

Clarity. The members of the Airmanship 2.0 Culture are clear about their intentions and understandings and they communicate and coordinate their clarity with the other members of the culture.

Alignment. Members of the Airmanship 2.0 Culture are continuously striving to remain in alignment with the other members of the culture.

Artifacts are the visible, tangible aspects of organizational culture. For example, a basic assumption employees and managers share might be that happy employees benefit their organizations. This might be translated into values such as egalitarianism, high-quality relationships, and having fun. The artifacts reflecting such values might be an executive open door policy, an office layout that includes open spaces and gathering areas and frequent company social gatherings. Some of the artifacts included in an Airmanship 2.0 Culture are:

Communications Channels. Emails, Twitter messages and phone calls to members regarding urgent safety issues must be used when necessary. Weekly and monthly eNewsletters as well as postings to blogs and Facebook and Linked In pages are used to maintain a robust communication between the members of the Airmanship 2.0 culture. Social networks (i.e., Facebook, Linked In, Twitter), social gatherings and one-on-one conversations between members serve as informal communications channels.

Awards and Recognition. Members of an Airmanship 2.0 Culture are recognized for their accomplishments on both an ad hoc and regular basis. The recognition comes in the form of publicity and awards.

Personal Airmanship Development Plan. Every member of an Airmanship 2.0 culture has a Personal Airmanship Development Plan that he/she has created with the assistance of the member's Mentor Pilot and Airmanship 2.0 training team.

Mentor Pilot Evaluations. Since it is very difficult for an individual to accurately assess his/her current level of airmanship knowledge and skill, a Mentor Pilot periodically evaluates members of an Airmanship 2.0 Culture to determine the member's airmanship strengths and weaknesses.

Shared Decision Making. An Airmanship 2.0 Culture includes policies, processes and procedures that support the making of shared decisions. The best example of this is the policy that requires that every flight be released with the concurrence of at least the flight's Pilot In Command and an FAA-certificated Flight Dispatcher.

Checks-and-Balances. Policies, processes and procedures that are designed to provide robust checks-and-balances are part of an Airmanship 2.0 Culture. For example, both the Pilot In Command and the designated crew chief for the flight check the fuel level before departure.

Backup systems. The availability of backup navigation, communications and flight-support systems are included in an Airmanship 2.0 Culture. All aircraft used by the members of the culture have hand-held navigation and communications radios and GPS navigators. iPads backup installed avionics.

Safety Education and Social Interaction. Frequent and accessible social interactions that feature safety education are woven into the fabric of an Airmanship 2.0 Culture.

Professional Support Team. Another artifact of the Airmanship 2.0 Culture is the structure, staffing, policies, procedures and quality of the professional team that manages and operates the business of the culture. The operational team must be closely aligned on the key components of the culture and they must exhibit behaviors that are in alignment with the standards, goals and expectations of the members of the culture.

Leadership. The leaders of the culture have to unambiguously and enthusiastically reinforce the culture's values, beliefs and assumptions through the policies and procedures they approve and through frequent contact with the other members of the culture. The leaders must consistently model the behaviors that are acceptable to the culture, and they must be vigilant for any signs that the culture's core values, beliefs and assumptions are being eroded by negative influences that have somehow crept into the culture.

This chapter has defined what Airmanship 2.0 is and why it was designed that way. Airmanship 2.0 has proven to be effective in radically improving airmanship whenever it has been wholeheartedly implemented. I firmly believe that personal flyers that opt to join an Airmanship 2.0 Culture will have a much safer, richer and more enjoyable personal-flying experience.

Chapter 3: *Airmanship 2.0 In Practice*

> *I learned that danger is relative, and that inexperience can be a magnifying glass.*
>
> —*Charles Lindbergh*

OK. You're seriously considering making a commitment to the pursuit of airmanship excellence through the practice of Airmanship 2.0. But what does it look like in practice and how do you find an Airmanship Development Support Organization (ADSO) that can support you in your pursuit? This chapter will answer those questions.

An Airmanship 2.0 Culture can come in many forms. In my opinion, two of those—airline and military—are the standards against which all other Airmanship 2.0 Cultures should be compared. In this chapter, I'll describe in detail what the pursuit of airmanship excellence can look like in practice within the context of an ADSO populated by personal flyers. However, please keep in mind that an ADSO can come in flavors other than the one depicted below as long as it adheres to the principles of Airmanship 2.0.

So, let me first answer the second question I posed above: How do you find a personal-flying organization that meets the criteria to be qualified as an ADSO? That answer is simple—you can't. Unfortunately, what we're going to talk about doesn't exist in reality at this time. It is currently at the "well-defined vision" stage. The business side of the ADSO described in this chapter has been extensively analyzed and there is a very competent team working on making it a reality. Nevertheless, as of now (Fall-2013), it doesn't exist. That means that I can't take you out to the airport and show it to you yet.

I can, however, answer the first question this chapter addresses—what does Airmanship 2.0 look like in practice—by sharing a story with you of what an ADSO can look like and how it can be created. The particular ADSO we're going to look at is structured in the form of a not-for-profit flying club. An ADSO can take several other business forms, but a properly structured flying club is the most stable and sustainable business model among them. To add a little realism to our discussion, let's call the ADSO in our story the "Palwaukee Aero Club" and base it at Chicago Executive Airport (ICAO identifier: KPWK) in Wheeling, Illinois. To better visualize this ADSO, let's set our story about it in the near future—the fall of 2016.

In an effort to provide even more clarity to our vision of how Airmanship 2.0 can facilitate your pursuit of airmanship excellence, let's look at the Palwaukee Aero Club through the eyes of three fictional people. Two of them have just decided to pursue their desire to learn to fly and one is a retired businessman and long-time personal flyer. All three of them live within a thirty-minute drive time from KPWK. Please keep in mind that these fictional but typical people represent only three of the many ways that people can discover, learn about and join an ADSO.

Our first avatar is a successful, non-pilot businessman who travels frequently for business-and-personal reasons. Most of that travel is within a radius of five hundred miles from his home in Deerfield, Illinois. He usually flies on the airlines only when the one-way drive time exceeds five hours. Otherwise, he drives his personal vehicle. He has, for some time now, harbored a desire to learn to fly so that he can use a personal aircraft for his business-and-personal-mobility needs. To make him easy to recognize, let's call him "Bill".

Our second virtual person is "John Jr.". He is a sophomore at Main South High School in Park Ridge, Illinois. He's been interested in a career as a professional pilot since he was ten years old. Unfortunately, he doesn't have anyone in his circle that can give him good advice on how to pursue his dream. Even the guidance counselors at his school know next to nothing about how to pursue a career as a professional pilot. They advised John Jr. to join the Air Force. That's not bad advice in itself, but it is only one of the many ways to become a professional pilot. Since he doesn't really want to enter military service, John Jr. has been left in the dark about how to get started towards the career he wants.

The third fictitious person in our group has recently retired from a very successful career as the founder and CEO of a mid-sized parts-manufacturing company located in Arlington Heights, Illinois. "Dan" served in the Vietnam War as an Army helicopter pilot. He didn't immediately continue to fly after returning home from military service because he couldn't afford it. However, a few years later when he founded his business, he joined a flying club to have access to a small, single-engine airplane. As Dan built his business, he used

Chapter 3: Airmanship 2.0 In Practice

that airplane, and a string of ever-more-capable general-aviation aircraft, to travel to his customers' factories.

These destinations were scattered around the Midwest, mostly in small towns with no airline service. Sometimes, Dan flew these airplanes himself, and at other times professional pilots employed by the company flew him and his executives to meetings. The company's sales team also used the aircraft to vastly increase their productivity.

Unfortunately, when he retired, Dan lost the use of the company's aircraft. For the past several months, he has been looking without success for a local flying club that can meet his rather exacting criteria for safety, support and community. Dan wants a selection of modern private aircraft available to him for his personal-mobility needs. He also wants to join a community of like-minded personal flyers, and he wants his personal flying to be supported by a professional flight-operations team.

Bill (the businessman who travels frequently for business-and-personal reasons) heard about the Palwaukee Aero Club from a golfing buddy who is a member of the club. Since both Bill and his friend share a similar need for business-and-personal mobility, and Bill has always wanted to learn to fly, the friend suggested to Bill that he look into the club as a solution to his business-and-personal mobility challenges. He also cautioned Bill that if he was going to learn to fly, he'd better do it right because his life and the lives of all his future passengers would depend on it. Since Bill is the kind of guy who always wants to do things right, he was intrigued by the challenge of proving he could learn to "fly right". Bill's golfing buddy invited him to attend an orientation briefing that was being held that weekend at the Palwaukee Aero Club clubhouse.

John Jr. became aware of the Palwaukee Aero Club through his association with the Main South High School Aviation Club. Early in his sophomore year (2016), driven by his frustration over not being able to find out how to pursue a career as a professional pilot, John Jr. did a Web search for youth-oriented aviation programs in his area. The National Association of High School Aviation Clubs (**NAHSAC.org**) was listed at the top of the first page of returns from his Web search.

John Jr. visited the NAHSAC Website and learned that the non-profit organization was funded and run by the Center For Airmanship Excellence (CFAE). NAHSAC was founded by CFAE as part of its Airmanship Outreach Program. CFAE trains, supports and compensates Airmanship Ambassadors who spend a considerable amount of their time meeting with local aviation and community groups and talking about Airmanship 2.0. CFAE Airmanship

153

Ambassadors also serve as the liaisons between high-school-aviation clubs who have joined NAHSAC and the local aviation community.

John Jr. was surprised to see on the NAHSAC Website that students in his very own high school were in the process of forming an aviation club. The contact information for the student organizers was provided. It didn't take long for John Jr. to join the group of students who were organizing the aviation club in his school.

The Aviation Ambassador assigned to help the group get the high-school-aviation club up and running trained the students in how to promote the new club within the student body and he provided them with the recruitment strategies and tools they needed to get things started. By late fall of 2016, the Main South High School Aviation Club had recruited fifty-five members and held its first meeting in a small auditorium at the school.

The meeting attendees included the students, their parents, the teacher-sponsor for the club and the Airmanship Ambassador. The Airmanship Ambassador briefed the attendees on the best ways the students could pursue professional-flying and other aviation careers. At the conclusion of the meeting, the Airmanship Ambassador invited John Jr. and his parents to the Palwaukee Aero Club clubhouse for an orientation briefing that was being held that weekend.

Dan (the retired CEO who had learned to fly in the Army during the Vietnam War) first heard about the Palwaukee Aero Club at an FAA Wings Program Airmanship 2.0 seminar that he attended. After the seminar, he did a Web search and discovered that the Palwaukee Aero Club had a modern, very-well-done Website. The site told him enough about the club to peak his interest. So, he called the club to see how he could go about learning more about it. The membership-services person he talked with on the phone invited him to attend a free orientation seminar that was being held at the Palwaukee Aero Club clubhouse that weekend.

Well, that's the background on the story I'm going to tell you about how Airmanship 2.0 can be put into practice. As you read it, I suggest that you think about whether or not the Palwaukee Aero Club is the type of personal-flying organization that you'd like to belong to. Enjoy the story.

The Future: Palwaukee Aero Club (0800 on a Saturday in the fall of 2016)

Bill, John Jr. and his parents, and Dan arrived bright and early on a Saturday morning for the orientation briefing at the Palwaukee Aero Club clubhouse. The clubhouse was located on Chicago Executive Airport. The clubhouse parking lot was secure and could be accessed only through a security-control point manned by club security personnel. Only club members and their designated guests are normally allowed to enter the grounds. However, Bill, John Jr. and his parents, and Dan were granted access because their names appeared on the approved-visitors list for that day. The security guard told them where to park and directed them to proceed to the main clubhouse entrance where a guide would greet them.

As John Jr. and his parents approached the entrance to the clubhouse from the parking lot, John Jr. remarked, "Wow, that's really a cool building." His parents nodded their heads and John Jr.'s father said, "This looks like a really first-class operation." John Jr.'s mother commented, "I can't wait to experience the inside of that beautiful building."

The main entrance to the clubhouse opened on a glass-enclosed, three-story, 100' x 100' atrium. The glass wall opposite the entryway looked out over a large ramp area where several private aircraft were parked under a canopy that extended from the clubhouse out two hundred feet from the building. Large balconies on the second and third floors of the atrium extended inward and outward from the three-story glass walls. Chairs and tables were arranged on the balconies that provided members and their guests with comfortable areas for casual meetings, quiet areas for studying and a spectacular 360° view of the airport and surrounding area. The three-story atrium was capped with a glass geodesic dome.

To the left of the entrance, the visitors could see the club's training center through a glass wall. The training center's ten **simulators** were all in use and other people were using **computer-training carrels** in another section of the training center. Four classrooms could be seen just beyond the simulator bay.

When the visitors turned to their right, the view through the glass wall of the atrium was of the club's operations center. Several people were working at computer workstations that were arranged in rows facing three large-format TVs mounted on the wall opposite the atrium. They were the FAA-certificated flight dispatchers and member-service representatives that provide flight-operations support to the club's members. The operations center looked out over the atrium and the ramp. When he took a closer look at the club's operations center, John Jr. was reminded of the pictures he had seen of NASA's Mission Control Center.

A hallway ran off to the right from the atrium along its parking-lot side. It's glass walls provided a view of the operations center on its left side and the parking lot and the clubhouse's well-manicured grounds on its right. The visitors could make out the entrance to the club's restaurant and lounge just beyond the entrance to the operations center. A similar hallway ran off to their left. The visitors could see the entrance to the club's business offices at the far end of this hallway beyond the training center.

The Chair of the club's Airmanship Outreach Committee, Diane, was standing in the middle of the atrium. She greeted Bill, John Jr., his parents and Dan as they completed their initial visual scan of the facility. After introductions were made all around and nametags distributed, Diane informed them that they would be joining a group of fifty other folks who were interested in finding out more about the Palwaukee Aero Club.

She then called over a middle-age man who she introduced as Arnie. He was the guide for Alpha Flight and one of the club's Mentor Pilots. She told all five of them that they would be members of Arnie's Alpha Flight for the day. Bill noticed that groups of ten were being formed and escorted towards the classrooms by friendly people dressed like Diane and Arnie in dark-blue Palwaukee Aero Club shirts and khaki pants.

Arnie led them to one of the classrooms in the training center where they met the other five members of Alpha Flight. Dan noticed that there were four other flights forming up in the classroom—Bravo Flight, Charlie Flight, Delta Flight and Echo Flight. After chatting for a few minutes with other members of their flight and helping themselves to some complimentary refreshments that were available in the back of the classroom, Bill, Dan, John Jr. and his parents took their assigned seats.

Diane stepped to the front of the classroom and opened the briefing by welcoming everyone again and introducing the five guides. She told the group that the orientation briefing would include a series of presentations made by her and the other guides. These presentations would cover the history of the club and show them how the club is structured. They would also be briefed on the club's fleet of aircraft. After these three briefings, the flights would separately tour the clubhouse and visit the hangars where the club's aircraft are stored. After the tour, they would be briefed on the club's training programs, how the club supports its members' pursuit of airmanship excellence and on how one becomes a member of the club.

Diane wrapped up the overview of the briefing by informing the group that after the last briefing, they would break for a complimentary working lunch in the club's restaurant. Each flight would sit together and each flight's guide would facilitate the generation of a list of questions from the members of the flight. These questions would be answered back in the classroom after lunch. She also asked the group to hold any questions about a particular briefing until the Q&A session after lunch. Diane then began her briefing on the history of the Palwaukee Aero Club.

Briefing 1: Palwaukee Aero Club History

"The Palwaukee Aero Club," Diane began, "was originally formed in 2013." "It started with a personal flyer named Steve who lives near Chicago Executive Airport. After reading *In Pursuit of Airmanship Excellence*, Steve was convinced that pursuing airmanship excellence through the practice of Airmanship 2.0 was for him. He wanted to have an Airmanship Development Support Organization—ADSO—to support him in that pursuit. He contacted the Center For Airmanship Excellence—CFAE—for help in finding an ADSO at an airport that was convenient for him."

"The then Chair of the CFAE Airmanship Outreach Committee responded to Steve's request with a phone call to him. The committee Chair told Steve there were currently two CFAE-certified ADSOs in the Chicago area—one at Chicago-DuPage Airport and one at Chicago-Aurora Municipal Airport. Unfortunately, these airports were over an hour's drive from Steve's home. So, they didn't meet his requirements for convenience."

"Steve then asked if the committee Chair knew of anyone who was interested in forming an ADSO at Chicago Executive Airport. After checking the CFAE database, the Chair told him that twenty-three pilots had notified CFAE that they would be interested in belonging to an ADSO at Chicago Executive Airport. However, no one had as yet stepped up to lead an effort to organize one there."

Diane continued, "The Chair explained to Steve that the Center was willing and able to help someone organize, set up and run an ADSO at Chicago Executive. But CFAE's policy was to only do that if there was a 'sparkplug' and a core group of interested pilots who were willing to act as the nucleus of the new organization. Steve told the Chair that he was interested in becoming the rallying point for a new ADSO at Chicago Executive."

"Steve was told that the next step would be to set up a meeting at CFAE Headquarters. The meeting would include Steve, nine other pilots who had put their names on the 'interested' list and the CFAE staff members who would be responsible for helping him make the ADSO a reality. Steve thought that was a great idea. It didn't take him long to recruit nine other people from the 'interested' list who wanted to join him for the meeting at CFAE Headquarters."

"During the meeting, CFAE staff briefed the participants on the options they had for going about setting up and running an ADSO. The staff also made some recommendations that took into consideration Steve's and the other personal flyers' capabilities and resources. By the end of the meeting, Steve and the nine other personal flyers had decided to move ahead with establishing an ADSO at Chicago Executive Airport."

"The following week," Diane went on, "with the help of a CFAE attorney, they registered the ADSO with the state of Illinois as a not-for-profit limited liability company. They named it the "Palwaukee Aero Club" in honor of the historical name for Chicago Executive Airport. Steve and the other nine personal flyers became the founding members of the club. They collectively loaned the club the modest capital it needed to cover necessary expenses related to the start up. Steve was unanimously elected as the first president of the club and he and four of the other personal flyers were appointed to the club's Board of Governors."

Diane continued, "After the club was registered with the state and the required start-up capital was in place, Steve, on behalf of the club, signed an ADSO Management Agreement with CFAE." "This agreement put the primary responsibility for establishing and running the club as an ADSO on CFAE. Since none of the founding members had any experience with setting up and running a flying club, this was a very attractive way for them to secure the benefits of an ADSO with the minimum of fuss and risk. CFAE would do all of the work necessary to launch and run the club, and the club's Board of Governors would oversee that work and make all of the critical decisions that would be necessary as the work progressed."

"As advertised, the CFAE ADSO Management Agreement included everything the club's founders needed to organize, structure, set-up, start-up and run the Palwaukee Aero Club. CFAE assigned one of their trained ADSO start-up specialists to oversee the delivery of the services included in the agreement and to act as the manager of the start-up process."

"CFAE took care of selecting suitable facilities for the Palwaukee Aero Club's interim clubhouse and the storage of the club's aircraft," Diane said. "The CFAE start-up team also arranged for the leases on the club's aircraft, simulators and other equipment, and they negotiated agreements with all of the vendors the club would need for its operations. And CFAE recruited, selected and trained the full-and-part time management and staff needed to keep the club running smoothly. In other words, CFAE took care of everything that was necessary to get the club up and running. All the founding members had to do was to provide oversight of the process on the club's behalf and assist CFAE with the recruitment of the critical mass of members needed to actually launch club operations."

"The ADSO Management Agreement also included everything that was needed to establish and maintain an Airmanship 2.0 Culture. Initial and recurrent training for the club's staff on how to sustain the culture on a day-to-day basis was also provided. Ongoing management-and-staff recruitment, selection and training were also included in the agreement as was the club's business-operations plan that they would use to keep things running smoothly, efficiently and safely. The agreement also included CFAE certification as a conforming ADSO. The certification process includes semi-annual inspections and periodic audits of the club's operations."

Diane continued, "The ADSO Management Agreement included the club's Flight Operations Manual and its flight-operations plan. It also included setting up our operations center. We call this our OpsCenter. You'll learn more about that later. The club also signed an Exclusive Airmanship Training Agreement with CFAE that includes the CFAE Mentor Pilot Program, the CFAE Instructor Pilot Program and the CFAE Airmanship Challenge Program. All of the airmanship-training programs that are needed to initially qualify and continuously re-qualify the club's members to fly the club's aircraft are included in the agreement. Under the terms of this agreement, CFAE also set up our training center."

"The CFAE Airmanship Outreach Committee—AOC—assisted Steve and his growing number of club volunteers with the club's membership drive. As new members joined the club, the AOC trained them in how to participate in the club's referral program by networking to potential new members. CFAE also set up and ran the club's Website and other online-media communications channels, and it provided the organization and management of the club's membership drive as part of the CFAE ADSO Management Agreement."

"CFAE also supported the Palwaukee Aero Club's membership drive through the Center's Airmanship Ambassador Program. Five CFAE Airmanship Ambassadors were assigned to give a series of Airmanship 2.0 presentations to all of the pilot and aviation-business organizations that are based at Chicago Executive Airport. The CFAE Airmanship Ambassadors also made Flying 2.0 presentations to a large number of civic and business organizations in the communities that are located within a thirty-minute drive time of the airport. These presentations explained how members of these organizations could use modern personal-flying for their business-and-personal-mobility needs."

Diana explained, "During the initial membership drive, the club members and the CFAE team all had a great time talking to potential new members." "It's always a pleasant experience when you can share something you believe in with people who are interested in hearing about it. At the outset of the membership drive, the Board of Governors had planned on a one-year campaign to reach the critical mass of fifty new members. In the end, it had only taken six months."

"There was quite a flurry of activity when the critical mass of fifty new members was reached. At this point, the members' initial membership fees were released from the escrow account they had been deposited in when they signed up. Now that there was adequate funding for the launch of the club, the CFAE start-up team took care of securing and building out office space for the club's interim clubhouse and training center. CFAE also arranged for the lease of an initial fleet of five aircraft and the five private, individual hangars at Chicago Executive Airport to store them in. The CFAE start-up team also arranged for the lease of three simulators (one for each type of aircraft in the club's fleet). The simulators were installed in the club's training center. CFAE also recruited, selected and trained the management and staff needed to run the club. All of these tasks were accomplished over the ninety-day period after the members' initial membership fees were released from escrow."

"Last year," Diane said with a note of pride in her voice, "we moved into this new clubhouse." "We're really enjoying its ambiance and amenities. As a member of the Palwaukee Aero Club, you'll have access to it 24/7."

"Now," Diane said, "I'd like to conclude my briefing by telling you about the rapid growth of the Palwaukee Aero Club since it was set up in 2013. At this point, three years after the club was founded, there are a little over three hundred members in the club. We're on track to reach our membership goal of five hundred members by the end of next year. And when we reach that goal, we won't take in any new members unless someone drops out of the club."

Diane then introduced Arnie, the guide for Alpha Flight. She told the group that Arnie's day job was flying as a captain for a major airline. He was an early member of the club, an FAA-certificated flight instructor and a CFAE-certified Mentor Pilot. Arnie, she told them, would be briefing them on how the club is structured.

As Arnie walked to the front of the classroom, Bill, who was seated next to Dan, leaned over and said to him, "Boy, it sure looks like a membership in this club is really a hot item." Dan turned to Bill and replied, "It sure does." "I guess we'd better snap one up before they close the membership."

Briefing 2: Palwaukee Aero Club Structure

After a short round of applause, Arnie began his briefing. He asked the group, "Do you know what makes airplanes fly?" The question elicited a few responses from the group like "the wings", "the engine", "the pilot" and "magic". Arnie told them, "Those are all good answers, but there is a more-essential factor that makes airplanes fly. And that's money." He then quoted Mercury Astronaut Gus Grissom from the movie "The Right Stuff". Arnie told them that the quote was, "No bucks, no **Buck Rogers**."

Chapter 3: Airmanship 2.0 In Practice

There were only a few chuckles around the classroom from those who understood the reference. So, Arnie went on, "Airmen have to be able to see things as they are and not rely on hope." "The cold, hard truth of the matter is that flying is, although affordable for most people who really want or need to do it, a relatively expensive pursuit. And to do it right, it even costs a little more." Arnie also shared with the group his belief that if you're going to fly, you need to do it right.

Arnie told them, "Since you are attending this orientation briefing, I know that you are considering joining the club." He cautioned them to do their best to become fully informed about the real cost of learning to fly. "And," Arnie continued, "You have to get clear on the true value of using a private airplane for business-and-personal mobility."

He recommended that they look into other ways to access private aircraft as well as the club. Arnie passed out a list of flying clubs, flight schools and aircraft-rental operations in the Chicago area and encouraged them to explore the value propositions they all offered. He told them, "By the end of today, you will have a very good idea of how we do things around here." "I'm sure you'll know the right things to look for and the right questions to ask when you do your due diligence by checking out these other options for personal mobility."

This "No bucks, no Buck Rogers" preamble led to an overview of how the club was structured. Arnie told the group, "The Palwaukee Aero Club was registered with the state of Illinois in 2013 as a non-profit Limited Liability Company (LLC)." "It qualifies with the IRS for 501 (c) (7) (Social Club) tax status. Members of the club can vote in periodic elections of the club's Board of Governors and officers. These elections are normally held every three years." "In fact," he told them, "there will be an election for the club's president and two of its governors this year."

Arnie continued his briefing by telling the group that CFAE had drawn on European aero clubs and U.S. flying clubs for the design of the business model that the Palwaukee Aero Club was using. CFAE's main goal was to design a business model that was sustainable, had little-to-no risk from a financial perspective and was easy to understand and administer. "The big advantage," Arnie told them, "the Palwaukee Aero Club business model has over the European-aero-club and U.S.-flying-club models is that it is staffed and run by aviation professionals rather than amateur volunteers."

"The Palwaukee Aero Club business model is pretty straightforward," he told them." "Members pay a Membership-Agreement Fee at the beginning of their five-year membership agreement. Then members purchase an Annual Membership Package that is based on their Personal Airmanship Development

Plan, the primary type of aircraft they will fly during the year and the number of hours they plan to fly that aircraft type during the year."

"Membership agreements have a five-year term and are renewable for five additional years with the option to renew for a total of fifty years. The initial Membership-Agreement Fee covers the member's share of the upfront funding on the leases for the aircraft, simulators, other equipment and the clubhouse. The Annual Membership Package fees collectively cover the lease payments and all of the club's overhead and operating costs during the year."

"Since the club actually owns no equipment or real estate—airplanes, simulators, clubhouse, etc.—but instead leases all of it, the lessors who lease the equipment and clubhouse to the club are of course very interested in the club's financial capability and stability. That's why all of the leases are structured on a five-year basis and the membership agreements run for five years. The lessors can be relatively secure in their expectation that the leases will be serviced properly. This security makes it easier for us to lease the aircraft and equipment we need, and it reduces our leasing costs. Our members also appreciate the financial stability that is designed into the club's business model."

"Another element of the business model that makes it low-risk is the fact that a new aircraft is not added to the club's fleet until there are at least ten new members who have committed to purchase five consecutive Annual Membership Packages on that type of aircraft. By adding aircraft to its fleet in this manner, the club doesn't end up with more aircraft than it can support. This approach also provides for well-planned and implemented expansion."

Arnie summed up his briefing by telling the group, "The business model the club has selected has demonstrated that it is sustainable over a long period of time, and almost all of the financial risk had been boiled out of it." "It is also easy for everyone involved to understand. The business model is relatively easy to administer since there are no accounts receivable and the operating budget automatically matches the operating plan with no potential deficits and no unfunded debt." At the completion of his briefing, Arnie told the group, "I sincerely hope that you acquire today the knowledge and wisdom you need to make a fully informed decision about whether or not you want to 'fly right' with the Palwaukee Aero Club."

Next, Diane introduced Linda, the guide for Bravo Flight. "Linda is the Chair of the Palwaukee Aero Club's Aircraft Selection and Maintenance Committee," Diane told the group, "and she is the chief pilot for a local corporate flight department." "Linda flies as a captain on her company's **Gulfstream G650** when she's not chained to her office chair with her administrative duties. In addition to being a professional pilot, Linda has a degree in aeronautical engineering and she is an FAA-certificated Aviation Maintenance Technician with Airworthiness Inspection authority. She is an FAA-certificated flight instructor and a CFAE-certified Mentor Pilot." "Linda," Diane told the group, "will be providing you with an overview of the club's fleet of aircraft."

As Linda took the podium, John Jr.'s father said to John Jr., "I like the look of this club." "They're certainly well organized and the club's business model appears to be rock-solid."

Briefing 3: Palwaukee Aero Club Fleet

Linda started out by asking the group if they knew what the Airmanship 1.0 emergency procedure is for an engine failure at night in a single-engine airplane. No one ventured an answer. So, Linda told them, "If the engine on your single-engine airplane fails at night, slide your seat back as far as it will go, bend over placing your head between your legs and then kiss your ass goodbye." At first, there were a few chuckles but they quickly died as the group realized that she wasn't kidding.

Linda continued, "In anything less than extremely lucky circumstances, your chances of surviving an off-airport forced landing at night are very slim at best. Since we all want to get maximum utility out of the aircraft we fly, we need to in many cases fly them after the sun goes down. But, how can we mitigate the risk of an engine failure at night in a single-engine airplane?"

Linda explained, "This procedure only applies to those pilots who continue to practice Airmanship 1.0 and fly single-engine airplanes at night." "Airmen who practice Airmanship 2.0, on the other hand, will not have to use that particular emergency procedure if the engine on their single-engine airplane fails on a night flight. And every member of the Palwaukee Aero Club practices Airmanship 2.0." "The practice of Airmanship 2.0," Linda told the group, "includes flying modern aircraft that are equipped with all available safety devices." "The way our club handles the risk of night single-engine flight is by equipping our aircraft with three safety devices that can significantly reduce the risks associated with an engine failure at night. These safety devices include a 'digital parachute system', a 'whole-aircraft-parachute system' and an 'enhanced vision system'."

"The club's Aircraft Selection and Maintenance Committee follows the recommendations of the CFAE Aircraft Selection and Maintenance Committee and it's Safety and Standards Committee. We therefore take great care to ensure that all of the club's single-engine aircraft are equipped with these three systems that can save our members' bacon if they are ever unfortunate enough to be aloft at night in a single-engine airplane with an engine that has just decided to pack it in for the night."

Linda put a slide up on the classroom's screen showing a picture of a 5" x 10" rectangular computer display with a control panel. The display was mounted on the instrument panel of one of the club's **Cirrus SR22s**. Linda explained, "This device is a backup Electronic Flight Instrument System — EFIS — that can be used to safely fly the airplane in the event of a complete failure of the primary EFIS. In this slide, you can also see the two large display-and-control screens of the **Garmin G1000 Perspective** navigation, guidance and control system installed in all of the club's SR22s. The two displays and associated controls make up the SR22's primary EFIS. We'll talk a little more about the wonders of the Garmin G1000 system in a little while."

"The backup EFIS is a **Vertical Power VP-400**. The VP-400 has several really neat features, but the one we want to talk about right now is its built-in 'Runway Seeker'. This nifty little device is always aware of the location of all the runways that your aircraft is capable of gliding to if the engine fails. If the engine does fail, the pilot simply pushes a button marked 'Runway Seeker' and the seeker selects the best runway that you can glide to, it provides guidance cues for you on the backup EFIS and it engages the autopilot to fly the airplane to a 'runway threshold' where the pilot can take over and land the airplane safely."

"That's why we refer to it as a digital-parachute system. For example, on a typical club flight, from the altitudes that the SR22 normally cruises at, there is at least one runway within engine-out gliding distance at least eighty-percent of the time. So, this means that most of the time, day or night, good weather or bad, your digital-parachute system can take you safely to a landing on a runway if you suffer the indignity of an engine failure."

Linda went on to tell them, "If the engine fails on one of the club's SR22s when it is four-hundred feet or more above the ground, and there is no runway within gliding distance, the pilot can deploy the **Cirrus Aircraft Parachute System** — CAPS. The pilot does this by simply pulling a handle that is mounted above his head. A small rocket instantaneously fires and pulls a parachute from its storage location behind the airplane's baggage compartment. The parachute quickly deploys and lowers the airplane gently to the ground. So far, there have been over one hundred successful deployments, as of 2016, of whole-aircraft parachute systems. Those deployments saved well over two hundred lives. They work!"

Linda told the group about the third safety device that is installed in all club aircraft that fly at night — an enhanced-vision system (EVS). "The enhanced-vision system installed on the SR22 is an **Astronics Max-Viz 600**

Chapter 3: Airmanship 2.0 In Practice

infrared-camera system," Linda said. She changed slides to show two images: one of the Max-Viz 600 infra-red camera mounted under the left wing of an SR22 and one showing an enhanced-vision display on the right screen of the G1000 in the cockpit.

"In the event of an engine failure at night in one of the club's SR22s," Linda told them, "the pilot can use the EVS to see the terrain below him and to see any obstacles that may be on his final-approach path to a runway that the Runway Seeker has taken him to. It also allows the pilot to see the runway and airport environment—even if the runway lights are not lit. And finally, the EVS can alert the pilot to any wildlife or livestock that is on the runway. These capabilities give the airman a distinct edge in surviving a 'deadstick' landing at night, whether it be on a runway or in an open field. Actually, the EVS is an important safety device on all night flights."

After explaining how these three safety devices, coupled with comprehensive initial and recurrent training, greatly improve the chances of walking away from a forced landing at night, Linda went on to tell the aspiring aviators, "But the best safety device you can have if your going to commit aviation at night in a single-engine airplane is a well-maintained engine." "So, let's talk about that." "To start off", Linda said, "I'd like to tell you about the things that a reciprocating engine, like those installed in our aircraft, needs to keep it running." "As we all learn very early in our airmanship training, those essential ingredients are fuel, spark, air and the mechanical integrity of the engine. So, your engine needs these four basic elements to keep on running whether it's day, night, good weather or bad. If your engine lacks any one of these constituents, it will stop running."

"Tackling these potential causes of engine failure one at a time," Linda continued, "let's talk about fuel first." "On our aircraft, there's an engine-driven fuel pump that moves fuel from the tanks in the wings to the engine whenever the engine is running. These pumps sometimes fail. So, our aircraft also have an electrically driven fuel pump that will continue to feed fuel to the engine in the event of an engine-driven-fuel-pump failure if the pilot turns it on with a switch in the cockpit."

"So, in our aircraft, about the only way that lack of fuel can cause the engine to quit is if you run out of it. You'd be surprised how many pilots who practice Airmanship 1.0 run their tanks dry or don't understand the fuel system in the aircraft they're flying well enough to properly manage what fuel they do have onboard. Every year, almost one-hundred Airmanship 1.0 pilots precipitate an engine failure with improper fuel management."

"As you've already heard, and you'll hear again several times today, every member of the Palwaukee Aero Club practices Airmanship 2.0 and the club supports that practice. Airmanship 1.0 pilots generally approach preflight planning in a rather casual manner, and they usually do all the flight planning by themselves. Poor preflight planning causes over fifty percent of personal-flying fuel-related accidents. Here's how Airmanship 2.0 handles the fuel-starvation hazard."

"All Palwaukee Aero Club flights are meticulously planned by a team that consists of an FAA-certificated Flight Dispatcher, a member-services representative, a senior dispatcher, a duty chief pilot and the airman. The dispatcher assigned to the flight and the airman who is the Pilot In Command—PIC—must agree on the flight plan, including the fuel load, before the flight can be dispatched. This process and policy reflect the club's shared Airmanship 2.0 values of checks-and-balances, teamwork and shared decision-making. They ensure that each flight departs with enough fuel to reach its intended destination with an appropriate reserve."

"But," Linda continued, "the Palwaukee Aero Club goes even further to keep any of its aircraft from running out of fuel." "Before departure, the PIC and the ground crew that is helping him get the airplane ready for the flight both verify that the fuel on the airplane matches the fuel load specified in the flight-dispatch release that the PIC and the dispatcher agreed to. Then, enroute the integrated avionics systems in our aircraft provide the PIC with very accurate fuel-onboard, fuel-remaining, fuel-endurance and fuel-range information. It's very easy for the PIC to see if he has enough fuel remaining to reach his destination plus an appropriate reserve."

"In addition to having all of this data at his fingertips, the enroute procedure calls for the PIC to call flight dispatch about half way through the flight on the aircrafts' satellite phone to report the amount of fuel onboard. The call is made to the dispatcher who is following his flight in real time. The dispatcher then independently verifies that the remaining fuel will get the flight to its destination with adequate safety reserves. If there is sufficient fuel onboard, the dispatcher and PIC re-dispatch the flight to its destination. By operating in this Airmanship 2.0 manner, we've never even come close to running out of fuel on a club flight."

"At this point," Linda said, "we've pretty much eliminated fuel starvation as a factor in the failure of one of our engines." "However, there are also the hazards of fuel contamination, fuel-filter clogging and broken fuel lines. The fuel-contamination issue is handled with rigorous preflight fuel-quality testing. The clogged fuel filter and broken-fuel-line risks are mitigated by more frequent and more-focused maintenance inspections."

Linda continued, "The second thing a reciprocating engine needs to run is a spark." "In a reciprocating aircraft engine, the spark is provided by two separate ignition systems. Each system has an engine-driven magneto. The magneto sends electricity to sparkplugs that are located in the engine's cylinders. There are two sparkplugs in each cylinder and there are six cylinders in the engine. Therefore, there are a total of twelve sparkplugs in the engine. As long as the propeller is rotating, the magnetos are producing electricity. The magnetos don't need electricity from the aircraft's battery or alternator. The chances of both engine-driven-magneto-ignition systems failing at the same time are very, very small. Here again, excellent maintenance ensures the reliability of the ignition systems."

"The third thing an engine needs to run is air," Linda told them. "The air the engine needs enters the engine through openings in the nose cowling. The engines in our aircraft are fuel injected, so we don't have to worry about carburetor icing blocking the flow of air to the engine. If the openings in the cowling become iced over or blocked by something like a large bird, there is an alternate air source for the engine. So, we really don't need to worry about our engines quitting due to lack of air."

"The fourth thing that I mentioned that a reciprocating engine needs to keep running is mechanical integrity. If a major engine component fails, then the engine will in all likelihood quit running. The key here is to make sure the engine is continuously monitored, and inspected and maintained frequently by a highly skilled Aircraft Maintenance Technician—AMT."

"The Palwaukee Aero Club changes the oil in the engines on all of our aircraft every fifty hours of engine time. This happens about every month on average for each airplane. When the oil is changed, the AMT thoroughly inspects the engine for defects. Every two hundred hours of engine time, an AMT inspects the engine's cylinders and valves with an **Endoscope**. If any serious defects are found, they are fixed or the engine is replaced. This approach to engine maintenance goes a long way to ensuring that the engines on our fleet of aircraft keep running."

Linda continued, "In addition to great engine maintenance, the performance of our engines is monitored by an onboard digital engine-data-monitoring system." "This data is uploaded to a server after every flight and analyzed by engine-trend-analysis software. If any of the engine's operating parameters have been exceeded, or if the trend data indicates a potential engine-integrity problem, the OpsCenter is alerted and appropriate maintenance action initiated."

"As we can see," Linda said, "the chances of an engine quitting on a club aircraft are pretty slim." "And if an engine in one of the club's single-engine aircraft does fail, we have systems and procedures that make sure the airplane's occupants return to earth safely. Our approach to mitigating the risks associated with an engine failure at night in a single-engine airplane enhances the business-and-personal mobility of our members while providing them with a level of safety Airmanship 1.0 pilots only dream about."

Linda went on to tell the group, "And all of the aircraft flown by the club have additional safety features that make them much safer than the typical aircraft flown by Airmanship 1.0 pilots." "These include crashworthy airplanes, computerized cockpits, auto-flight systems, satellite-tracking systems, satellite phones, angle-of-attack indicators, head-up displays and airbags. Let's take a look at each of these features."

"The business-and-personal-mobility aircraft flown by club members include the Cirrus SR22 and **SR20** and the Flight Design **CTLSi**. All three of these aircraft are made from composite materials that are stronger, yet lighter than the conventional metal construction of most of the aircraft flown by pilots who practice Airmanship 1.0. This additional strength translates into a higher chance of survival in a crash situation. In addition, these aircraft have other crashworthiness features built into them."

"These three aircraft also have state-of-the-art computerized cockpits. The SR22 and SR20 use the Garmin G1000 Perspective Electronic Flight Instrument System that is specially designed for Cirrus Aircraft. The system includes two large displays, a keyboard controller, dual internal GPS navigators, a digital autopilot and a dual solid-state attitude-and-heading reference system—AHRS."

"The G1000's left display is called the Primary Flight Display—P-F-D. It is mounted in the instrument panel directly in front of the pilot. It shows a large-format, computer-generated synthetic view of the sky and the terrain out in front of the airplane. Vertical tapes show airspeed and altitude. Heading and course information, a moving map, traffic and other important information can also be shown on the PFD. In other words, all of the information a pilot needs to safely fly the airplane is presented to him on the PFD."

"The G1000's right display is called the Multi-Function Display or M-F-D. Just as its name implies, many different functions can be displayed on it. These include electronic checklists, flight plans, navigation and approach charts, moving maps, terrain-proximity maps, lightning-detection displays, NEXRAD weather-radar displays, engine-and-systems information, communications-and-navigation information and traffic information."

"The Flight Design CTLSi is equipped with a **Garmin G3X** Electronic Flight Instrument System. It provides the same basic functionality as the G1000 system. However, it uses a separate **Garmin GTN 750** GPS navigator. The CTLSi also has a digital autopilot."

Chapter 3: Airmanship 2.0 In Practice

"These 'glass cockpits' as we call them," Linda explained, "make personal flying a lot easier if the pilot using them has been properly trained on how to operate them." "You'll get a briefing on the club's training programs later today. But for now, let me assure you that every Palwaukee Aero Club airman knows how to best utilize the glass cockpits in the club's aircraft. We believe that glass cockpits can make personal flying a lot safer if they are used properly."

"I'm sure you noticed that I mentioned the fact that autopilots are installed in the Cirrus SR22, SR20 and Flight Design CTLSi," Linda said. "That's because in certain situations, auto-flight is safer and more convenient than hand flying the airplane. During auto-flight operations, the Pilot In Command is still in control of what the aircraft does. Only instead of moving the control stick by hand to maneuver the airplane, the PIC uses inputs to the flight computers to do that. These autopilots will capture and track navigation courses and even fly the airplane down a glideslope. And they have 'heading-hold' and 'altitude-hold' functions. Again all of our pilots are taught when to, and when not to, use these auto-flight systems."

"All of the club's aircraft are equipped with a **Spidertracks Spider S3** satellite tracking-and-communications system. A flight dispatcher in the club's OpsCenter is assigned to monitor the progress of every club flight. We call this 'flight following'. The Spider S3 in the airplane sends a GPS-position report to the OpsCenter through the Iridium Satellite Network in real time. The system operates anywhere in the world at any altitude. The flight dispatcher follows the progress of the flight on a computer display."

"If the pilot experiences an emergency like an engine failure that is going to cause a forced landing, he can press an 'SOS' button on the Spider S3 control panel and an email message is sent directly to the flight dispatcher in the club's OpsCenter. The dispatcher will immediately contact search-and-rescue responders who can then proceed to the exact location where the aircraft landed. This takes the 'search' out of 'search and rescue' and gets a rescue team moving to the proper location much faster than would normally be possible."

"Since the flight dispatcher is constantly monitoring a flight's position, the dispatcher can back the pilot up in terms of adverse weather and other hazards that the pilot will want to be aware of. In this event, the dispatcher will call the pilot on the aircraft's satellite phone to talk with him about the hazard. The satellite phone itself is an important safety device that Airmanship 1.0 aviators don't usually have at their disposal. I particularly like this feature myself because I never feel like I'm out there all alone." "It significantly increases my comfort level when I'm flying club aircraft," Linda told them. "I use it whenever I want assistance in gathering information, changing my plans enroute or to get help with any problems I may be encountering. And I really like the fact that the dispatcher following my flight can call me if there is any new information that he thinks I should know about."

Linda continued her briefing by saying, "Another safety device that makes flying club aircraft a lot safer is the **Angle-Of-Attack Indicator** that is installed in every club airplane." "Angle of Attack is the angle between the wing and the air through which it is moving. One of the leading causes of fatal personal-flying accidents is loss of control caused by the pilot trying to fly the aircraft at an angle of attack that is too high to allow the wing to continue to produce adequate lift."

"That happens because without an angle-of-attack indicator, the pilot has to approximate the angle of attack by looking at his airspeed indicator. The problem is that the airplane's weight and dynamic loading affect the airspeed at which an aircraft's wing reaches what we call a 'critical angle of attack' and loses adequate lift. For example, an airplane can reach its critical angle of attack at fifty miles per hour when traveling straight ahead, but that speed may be seventy miles per hour in a steep turn. With an angle-of-attack indicator, the pilot always knows what the wing's angle of attack is. He doesn't have to guess at it. This makes maneuvering flight close to the ground much safer."

"We also equip our aircraft with **Head-Up Displays** or H-U-Ds. A HUD consists of a transparent display called a 'combiner glass' that presents flight data in the pilot's normal line of sight. The name comes from the fact that a pilot doesn't have to 'look down' at his flight instruments during critical phases of flight. HUDs were initially developed for military aviation, and they have been in use by the airlines and some corporate-aircraft operators for many years. HUDs are now available for the types of aircraft the club flies. A HUD makes personal flying much safer during relatively risky takeoff and landing maneuvers."

"The last piece of safety equipment I'm going to brief you on today," Linda said, "is the **Airbags** that are installed in all of our club aircraft. When sensors detect a sudden stop, the airbags deploy up and away from a seated pilot or passenger. They provide protection to the head, neck and torso. Aircraft airbags have been credited with saving lives and reducing injuries in aircraft accidents of all types. Pilots who practice Airmanship 1.0 don't usually fly airplanes that are equipped with airbags."

"Now I'd like to give you a quick overview of the aircraft we have in our fleet. Let's start with the Cirrus SR22. I fly it all of the time for my business- and-personal mobility needs, and I love it. In my opinion, it is the best piston-single-engine personal aircraft ever built." "It has all of the safety features that are available for that class of aircraft and it has a lot of neat stuff to make flying a considerably easier and more pleasurable experience for our members," Linda said.

"The Cirrus SR22 is one of the fastest piston-single-engine airplane in its class — 185 knots, that's 213 miles per hour. It needs only a little more than one thousand feet of runway to get into the air and it lands in about the same distance. It climbs at over 1,400 feet per minute and it can cruise at 17,500 feet. It has a maximum range of over one thousand miles. It normally seats four large adults, but it can seat five if three of the passengers are on the small side. It also carries luggage in a baggage compartment that's located behind the rear seats."

"The SR22 is also equipped with an aircraft-deicing system that clears ice off the leading edges of the wings, propeller tail and windshield in flight. This capability is critical to all-weather-flying safety. Most of the aircraft flown by Airmanship 1.0 pilots do not have this capability. There are additional safety features incorporated into the SR22's design that I don't have the time to go into right now. Let me sum up by saying that the SR22 has won the Robb Report 'Best of the Best' award more times that any other aircraft in any category."

"The Cirrus SR20 is really the same airplane as the SR22, but with a smaller engine. The SR22's engine produces 310 horsepower and the SR20's puts out two hundred. The SR20's maximum cruise speed is 178 miles per hour. It needs a little more runway to takeoff on than the SR22, but in lands in about nine hundred feet. The SR20 is less expensive to operate than the SR22. So, some of our members use it for their business-and-personal-mobility needs when they don't require the speed of the SR22. We also use it in our flight training program — more on that later."

"The Flight Design CTLSi is a two-seat, high-wing, tricycle-gear 'Light Sport Aircraft' or L-S-A. It is the best selling LSA in the world. Flight Design, a German company, has been designing and building airplanes since the 1980s and the first airplane in the CT series first flew in 1996. The CTLSi's safety record is very impressive and it is very economical to operate. It cruises at 132 miles per hour, but its one hundred horsepower engine consumes less that four gallons of fuel per hour. This means the CTLSi will get you where you're going at almost three times the average speed of auto travel at a very efficient thirty-three miles per gallon. Our members use it for their business-and-personal-mobility needs and for pleasure flying around the local area."

"For our members who want the increased safety, reliability and capability of a twin-engine airplane, we added the **Diamond DA-42 Twinstar** to our fleet," Linda said. "It is the most technologically advanced twin-piston-engine-powered airplane in the world. Its airframe is made from super-strong composite materials and its two Austrian-built diesel engines give it great performance. It can easily takeoff or land on a 2,000-foot runway and it cruises at 162 knots—that's 186 miles per hour. It has a full-fuel range of almost one thousand miles and it's equipped with the Garmin G1000 system."

"The SR22, SR20, CTLSi and Diamond DA42 Twinstar form the core of our fleet. However, we also have other aircraft that our members use for other kinds of flying. For example, we have a couple of **AirCam** twin-piston-engine aircraft that our members use for sport flying. Its two-seat, open-cockpit design coupled with its two-engine reliability and safety, make it the perfect platform for aerial photography, exploring the countryside or just enjoying a warm summer evening of pleasure flying. It's very easy to fly and relatively inexpensive to operate. One of our AirCams is equipped with **amphibious floats**."

"We also have," Linda told them, "a **Waco YMF-5D Super** open-cockpit biplane." "It's a new version of an airplane that was first designed in the 1920s. It has two open cockpits, two wings and a tailwheel. It's the kind of airplane the old barnstormers used to fly. Only the Super has been upgraded with modern safety features and modern avionics. It's a fun airplane to fly. Many of our members use it to give rides to their family and friends, and to just tool around on nice sunny days."

"Speaking of modern versions of old airplanes, we have several Cubs in our fleet. Our members use a modern version of the venerable **Piper J3 Cub** for their initial and recurrent training and for certain Airmanship Challenges. The modern Cub that we use is the **CubCrafters Sport Cub S2**. You'll learn more about the Cub later today when you're briefed on our airmanship-training programs."

"In the future, we plan to add additional aircraft to our fleet. The types of aircraft we will add depends on recommendations from the Center For Airmanship Excellence and what our members want to fly. If we have enough members who are willing to commit to collectively fly a type of aircraft a minimum of four hundred hours a year, we can justify adding it to our fleet."

"Well, that completes my briefing," Linda said. "If you have any questions, please jot them down and we'll answer them in the Q&A session after lunch. Now, let me turn the briefing back over to Diane. Thank you very much for your attention."

The group gave Linda an enthusiastic round of applause as Diane moved to the front of the classroom. John Jr. leaned over and said to his mother, "See Mom, I told you these airplanes are really safe to fly." His mother replied, "Yes son, I'm beginning to see that what you've been telling me about how safe flying can be is true." "In fact, these airplanes look so neat to fly that I may take flying lessons myself," she said.

When the applause died down, Diane said, "Now we're going to break the group down into our five flights. Your guide will take you on a tour of our clubhouse and the hangars where we keep our aircraft. After the tour, we'll rendezvous back here in this classroom for three more short briefings before we go to lunch. Enjoy your tour."

The Tour: OpsCenter

Arnie rounded up the ten members of Alpha Flight and directed them to meet him in the clubhouse's atrium after a ten-minute break. After Alpha Flight reassembled near the atrium's entrance, Arnie led them to the viewing room overlooking the club's OpsCenter. When Alpha Flight was in the viewing room, he told them, "This is our operations center, or as we like to call it our 'OpsCenter'." "It plans and coordinates all of our flight activities."

Arnie continued, "As you can see, there are about ten people working here on this shift." "The OpsCenter is staffed twenty-four hours a day, every day of the year. Each OpsCenter staff member has his or her own workstation. It's equipped with three monitors, a computer and telephone equipment. You can see the three large monitors mounted on the front wall. They display general information and maps that are used by everyone in the room."

The members of Alpha Flight noted that the big screen in the center displayed a map of the United States with several airplane icons. Attached to each icon, a data block identified the airplane icon as a club aircraft and provided pertinent flight information. The screen on the left showed another map of the U.S. with weather symbols and radar returns displayed on it, and the big screen on the right showed a table that contained the registration numbers of all of the club's aircraft in the left column of the table. Other columns displayed additional information about the type and status of each airplane.

As the members of Alpha Flight took in the scene, Arnie resumed his briefing. "But those displays, along with several others, can also be called up on the individual monitors at each workstation. You'll note that the OpsCenter team can look out the large windows on their left at the club's ramp and the airport beyond. That helps them to stay grounded in the real world while they spend their shift mostly in a virtual reality."

"The primary mission of the OpsCenter team is to support our members before, during and after a flight," Arnie told them. "The team is composed of flight dispatchers, member-service representatives, a lead dispatcher and a duty chief pilot." Arnie informed them. "The team can also bring the duty maintenance supervisor into the loop via an online meeting or telephone hookup when his expertise is needed. In fact, the OpsCenter can set up an online meeting or conference call that can provide our airmen with any help and support they may need during the pre-flight, enroute and post-flight segments of their flights."

"In flight, hat's done via a satellite-phone link between the aircraft and the OpsCenter. I see that the dispatcher at workstation five is talking on his satellite phone link right now. I can tell because the 'satellite-phone-in-use icon' is displayed in the upper left corner of his center monitor. The dispatcher's name is 'Mike'. I'll connect the speakers in the ceiling over our heads with Mike's workstation so that we can hear his conversation with the airman." With that, Arnie tapped a few keys on a keypad located just below the main viewing window. Everyone in the viewing area could hear the conversation between Mike and the pilot.

"Roger six-six-kilo," Mike said. "I see that line of thunderstorms you're looking at. They're about fifty miles from your destination airport and moving towards it at about thirty knots. It looks to me like they're going to get there before you do. Please standby one and I'll bring the lead dispatcher and duty chief pilot into the loop." Mike tickled his keyboard and Alpha Flight soon heard both of them join the satellite-phone link with the pilot.

There was a brief discussion of the airman's options and the implications of deciding to execute any of the available alternatives. In short order, the pilot and Mike agreed that the best course of action was for the airman to divert to an airport that was only a short distance from his current location. The plan was for him to land at that airport, put the airplane in a hangar and wait for the thunderstorms to pass the destination airport and clear the pilot's course to it.

Arnie cancelled the communications link to the viewing-room speakers and said, "What's going to happen now is the pilot will talk with Air Traffic Control and get a clearance to the diversion airport." "Mike will call the fixed-base operator at the diversion airport and arrange for a hangar to put the airplane in until the storms pass. And he'll order appropriate servicing for the airplane that is based on the flightplan he'll put together from the diversion airport to the intended destination airport. While he's dong that, a member-service representative will set up ground transportation and dinner reservations for the airman at the diversion airport and call the fixed-base operator at the intended destination airport to let them know that the plan has changed and to expect the arrival of our member later today after the storms pass."

"The beauty of this OpsCenter-support system is that the airman's work load has been drastically reduced. This is a very important factor in flight safety. Also, a safer and more efficient alternate plan was arrived at and executed by the team than the pilot could have put together in the cockpit by himself. It is quite a challenge to re-plan a flight all by yourself while you're dealing with a highly dynamic flight environment. In fact, this is exactly where many Airmanship 1.0 pilots get themselves into trouble."

"In addition to this inflight support," Arnie said, "the OpsCenter team does several other things to make all of the flights flown by or for our members safer, more efficient and easier." "For example, it takes a good hour to put together a well-researched and thought-out flightplan. That's OK, but our members are busy people. It's not a good idea to put the burden of developing a good flightplan solely on their shoulders. We all have a tendency to rush things when we're in a hurry. That's not conducive to good flight planning or to aviation safety. So, the OpsCenter team takes care of this task. It results in a much safer and more efficient flightplan and a lot less work and hassle for our members. Of course, the member has the final authority to decide if a particular flightplan is used for his flight. This is the same system we use at the airline I fly for."

"Also, the member-service representatives in the OpsCenter lessen the burden on our members by taking care of all of the ancillary travel arrangements associated with a flight in a club airplane. They make hotel reservations, arrange for ground transportation and catering; and they will take care of anything else that a member may need like dinner reservations, event tickets and suggested tours of local attractions."

"So, for example, I can land at an airport and simply park the airplane. If I've specified that I want a rental car standing by, it will be parked right next to where I park the airplane. The ground crew will remove by bags from the airplane and put them in the rental car. In the car, there will be a map showing me where my motel is and a folder containing the tickets for a ballgame that I requested, and any other information that is pertinent to my stay. I can rest assured that the airplane will be taken care of properly by the ground crew and that it will be serviced properly for my departure. This type of support saves me a lot of time and hassles when I'm traveling by club aircraft."

"The OpsCenter also handles aircraft and pilot scheduling and positioning. Let me give you an example. Let's say a member is at his summer home near Two Harbors, Minnesota and he wants to schedule an SR22 to be available to him at the Helgeson Airport, that's the small general-aviation airport that serves Two Harbors, at eight-AM on Monday morning so that he can fly back to Chicago Executive Airport for a lunch meeting in Lincolnshire. He will be traveling alone and he's fully qualified to fly the SR22 by himself."

"On the Saturday before the planned Monday flight, our member submits a travel request to the OpsCenter. He can do that in several ways including by handheld device, computer or over the telephone. His travel request initiates a chain of events that will result in an SR22 sitting on the ramp at Helgeson on Monday at seven AM, fueled and ready to go with his flightplan filed and all other necessary arrangements for the flight completed."

"The first thing one of the member-service representatives in the OpsCenter will do is to ensure that one of the club's SR22s will be available for the flight. Then a dispatcher will check the forecasted weather for Monday and the airman's current qualifications to handle that weather. In this case, the weather is good and the member is qualified to handle it. The dispatcher will then create any flightplans that may be needed to service the member's travel request. In this case, the SR22 will be ferried to Helgeson Airport by a Center For Airmanship Excellence-certified professional pilot who will fly back to Chicago Executive with the member and assist him with the flight in any way the member desires."

Arnie continued, "Next, a member-service representative will schedule the airplane and the ferry pilot to fly to Two Harbors on Sunday evening so that the airplane is in position for the member's planned eight-AM departure on Monday." "The member-service representative will arrange to have the airplane put in a secure hangar overnight and order the planned aircraft servicing for the Monday morning departure."

"On Sunday the OpsCenter will be monitoring the weather for the flight on Monday. If it looks like it might deteriorate to the point where the member is not qualified to fly in it, the OpsCenter will inform the member that the ferry pilot will be the Pilot In Command for the flight back to Chicago. On the other hand, if the weather looks like it's going to hold up, the ferry flight will be released from Chicago Executive to Helgeson Airport and the member will be the Pilot In Command for the return flight."

"Now," Arnie said, "this is only one example of many that I could use to illustrate to you how the OpsCenter keeps everything moving safely, efficiently and smoothly. For example, the OpsCenter schedules all of our airplanes into maintenance and makes sure that they are airworthy and ready to go when they return from maintenance. And the OpsCenter makes sure all of our members' qualifications are up-to-date so that our members aren't burdened with keeping up with all that paperwork."

Dan, who was standing next to Bill in the OpsCenter viewing center turned to Bill and said, "Wow!" "This is exactly the kind of support that I had from my company's corporate flight department. I've been looking for a flying club that can provide me with it, but I haven't been able to find one. I think I have now. I believe I've found a home."

The Tour: Training Center

"Now," Arnie said, "if you'll follow me, let's go tour the training center." With that, Alpha Flight followed Arnie across the atrium to the training center. When they got inside, Arnie told them, "As you can see, we have simulators for each type of aircraft we fly." "There's an open sim over there. So, let's go over and take a closer look." Alpha Flight gathered around a sim that had *"Cirrus SR22"* painted on the outside just below the cockpit window. Arnie selected two members of Alpha Flight and invited them to sit in the two seats in the cockpit. Arnie sat down at the sim's instructor station and fired it up.

It wasn't long before the sim's visual system was projecting a computer-generated view of the ramp outside the clubhouse on a large, wrap-around screen. The visual scene was very realistic. When you looked out the sim's front windshield, it appeared just like it would if you were sitting in a real SR22 on the clubhouse's ramp. The G1000 avionics system on the instrument panel lit up like an arcade game. Arnie made a few entries on the sim's control panel and suddenly the simulated visual scene switched to a view of the airplane flying along at three thousand feet over a countryside dotted with lakes.

Arnie then failed the SR22's engine and walked the person sitting in the left seat through the whole-aircraft-parachute-system-deployment procedure. When the handle was pulled to deploy the 'chute, the visual scene showed the airplane's nose pitching up and then it showed the aircraft swinging under the parachute's canopy. The simulation continued a slow descent to the ground. When the simulated airplane hit the ground, the visual scene shook to simulate the touchdown. Murmurs of amazement and approval spread throughout the members of Alpha Flight.

After the comments died down, Arnie told the group, "We use these simulators for initial and recurrent training and for some of our Airmanship Challenges." "You'll find out more about them later. Also, our members can use these simulators and the other equipment in the training center anytime they want free of charge as long as no formal training is scheduled in them when they want to use it. Now, if you'll follow me, let's take a look at one of the computer-training carrels we have here in the training center."

The Flight walked a short distance to another glass-walled room that had about a dozen low-profile cubicles in it. Each cubical had a computer workstation and **Desktop Simulator** in it. "Our members use these training carrels for some of their training and to stay sharp on certain procedures," Arnie told them. "Every member receives one of these Desktop Simulators as part of his or her initial-membership package and they install them either at home or at their office. However, since we all live within a thirty-minute drive time of this training center, some us prefer to come here when we need a training carrel because there's always a Center For Airmanship Excellence instructor available to answer questions and help our members with the simulations."

"The only other thing we need to talk about here in the training center is the classrooms. You've already spent some time in one of them. There are three others, each of them seats fifty people. They're all equipped with the latest classroom-learning tools and of course they're wired. When we want to get a larger group together in the same room, we can open up the connecting walls between the rooms. That way we can seat over two hundred people in one large classroom. We keep the classroom pretty busy with various ground-school classes." "Now, if you'll follow me, I'll show you the club's business offices," Arnie said.

With obvious enthusiasm showing clearly on his face, John Jr. said loud enough for everyone in Alpha Flight to hear, "I can't wait to get my hands on one of those simulators!" Arnie replied, "Don't worry, you'll get all the time in 'The Box', as we call it, that you can handle and probably then some." "These simulators are incredible training devices, and the simulated real-life challenges that we can present in them will appeal especially to you John Jr."

The Tour: Business Offices

Alpha Flight left the training center, turned right and walked down a long hallway. On their left, they had a view of the parking lot and the grounds surrounding the clubhouse. On their right, they saw two of the four high-tech classrooms in the training center. At the end of the hallway, they passed through double-glass doors and assembled in the lobby of the club's business offices. John Jr. and his parents commented on how attractive the office space was—really professional. Just then, a man wearing a shirt with a Palwaukee Aero Club logo on it (just like the one Arnie was wearing) walked into the lobby and introduced himself.

"Hi everyone. I'm Steve and I'm the current president of the Palwaukee Aero Club. I'd like to welcome you all to our club and I hope you find your time with us today to be both informative and enjoyable. I'm sure Arnie is taking good care of you. In just a minute, I'm going to escort you down this hallway behind me to a door that leads to our aircraft-parking ramp. As we pass down the hallway, you'll get a good look at the workstations and offices here in our business office. We're not going to stop to talk with any of the people you'll see because I don't want to distract them from the important work they're doing for the club. But, they and I will be joining you for lunch. So, you'll get a chance to get to know them then. Now, if you'll please follow me?"

Steve led Alpha Flight down the hallway towards the airport-side door. Bill commented to Dan that it looked like everyone was busy. Dan added that he saw a lot of apparently happy people working in the business offices. Before long, they arrived at the end of the hallway.

The Tour: Hangars

The group could see a **Mercedes-Benz Sprinter** passenger van painted in the club's colors with its large passenger-entry door open invitingly sitting just outside the door. As Alpha Flight filed through the exit door and into the van, Steve shook each one of their hands and told them that he was looking forward to seeing them again at lunch. Upon entering the van, Bill noted that it would accommodate ten passengers in an executive-jet seating configuration. "Really first class," he whispered to Bill. Once seated, the attention of the members of Alpha Flight was drawn to the view out the left side of the van. They could see several club aircraft parked on the clubhouse ramp. Many of them were being loaded or unloaded by people who looked like they were enjoying the experience. Men and women in uniforms that identified them as members of the club's ground crew were attending to these folks.

When everyone was seated, Arnie stepped into the van and said, "As you can see, there's no room for me in the van." "So, I'll be taking that ground-support vehicle you see over there." The group looked where he was pointing and saw what appeared to be a golf cart made up to look line a miniature airplane. "I'll meet you at the hangars," he said. "And oh by the way, would anyone care to join me on the ride over. I can take one passenger." Several hands shot into the air, but John Jr.'s was the first one Arnie noticed. "If it's alright with your parents John," Arnie said, "why don't you join me on the ride to the hangar. I'll also be able to take one of you back from the hangars. So, to make it a little more interesting, when we finish our tour of the hangars, I'll ask the group an aviation related question and the first one to answer it correctly wins the return ride."

With that, Arnie and John Jr. headed for the golf cart. The van's driver closed the door and started moving towards the club's hangars. The driver came on the PA system and said, "Welcome aboard." "I'm Jack and I'm a ground-crew-team leader for the Palwaukee Aero Club. I'm a student-member of the club and I work for the club part-time. When I'm not helping the club's members to enjoy their personal-flying experience, I participate full-time in the club's Professional Pilot Preparation Program. I'm currently qualified to fly all of the club's aircraft and I'm working on earning my professional-pilot credentials. You'll hear more about the Professional Pilot Preparation Program in your next briefing." As he slowly maneuvered the van towards four buildings, Jack continued, "If you'll look out in front of the bus, you'll see four long, low buildings that are painted in the club's colors." "You'll note that the buildings have six big doors along each side and that each big door has a little door set into it."

"Those buildings are the club's **T-hangars** where the club's aircraft are secured. We call them T-hangars because the internal arrangement of the buildings is laid out in interlocking 'T' shapes to maximize the use of the space. You'll see what I mean in just a minute. The club currently has a total of forty-eight aircraft parked in those hangars."

Meanwhile, Arnie and John Jr. carried on a conversation while they "flew" in an in-trail position on the van. It started when Arnie asked John Jr., "So, why are you here today John?" John Jr. replied, "I really, really want to become a professional pilot like you and I want to find out how the club can help me do that." "Well," Arnie said, "you've certainly come to the right place." "I can highly recommend the club's Professional Pilot Preparation Program to you. You'll be hearing a lot more about it in the briefing we'll attend right after the tour of the hangars. And you'll get all your questions answered and concerns addressed after lunch."

As the van and the golf cart approached the hangars, the large door on the closest hangar began to open. As it did so, a gleaming Cirrus SR22 was revealed. Arnie pulled ahead of the van and arrived at the hangar first. He parked just inside the hangar. Arnie and John Jr. got out of the golf cart and Arnie beckoned to the group in the van to join him and John Jr. in the hangar. When everyone was assembled in front of the SR22, Arnie said, "This is one of forty-eight virtually identical hangars in this complex." "As you can see, the floor has a highly polished white surface that Jack and the other members of our ground crew keep looking like new at all times."

Dan said loud enough for everyone to hear, "That floor looks so clean you could eat off it." Arnie replied, "You probably could, although I certainly wouldn't recommend it." When the chuckles died down, Arnie continued, "You'll also note that the finished walls are painted white and various items of safety and ground-operations equipment are mounted to the walls and clearly labeled. As you can see, the hangars are kept clean, organized and properly equipped by our ground crews."

"If you look directly above the SR22, you'll see that a video camera is mounted on the hangar ceiling. There are also cameras mounted strategically around the hangar that afford a view of all four sides of the airplane. We use those cameras for several purposes."

"First of all, they are part of our Safety Management System. If there is a mishap in the hangar, like an airplane sustaining damage—we call it 'hangar rash'—while it's being put away, we use the recorded video to help us analyze how the accident happened. Then we can figure out how to mitigate the risk of it happening again and communicate the solution to our members in various ways. This saves the club money in the long run and it makes our operation that much safer."

"I'd like to point out that we don't use the recordings to assign blame," Arnie said. "There is no blame in an accident. That's exactly what it is—an accident. No one intentionally caused the mishap. Our Airmanship 2.0 culture fosters an atmosphere of non-punitive error recognition. We never blame or punish a member for an honest mistake, and it is in all of our interests to share the errors we make so that we can all hopefully avoid them in the future."

"The cameras you see mounted around the hangar, as well as the wireless, hand-held camera you see mounted on the wall just outboard of the SR22's left wingtip, are used to assist our members in their preflight inspections. If they find a mechanical discrepancy that they think might render an airplane unairworthy, the OpsCenter can set up an online meeting that includes the member, a flight dispatcher, the chief pilot and our maintenance supervisor. This team then uses the cameras to help the member make a decision on whether or not to fly the airplane. This capability serves our Airmanship 2.0 cultural values of 'shared decision making', 'checks and balances' and 'teamwork'. The cameras also support our initial and recurrent training programs and they provide an easy way for our Mentor Pilots to evaluate our members' proficiency in preflighting an airplane."

"We also use the cameras as part of our asset-security program. The OpsCenter and security software monitor the video feeds from the hangar to ensure no one is in the hangar when they're not supposed to be and that nothing is being done to the airplane. As I'm sure you noticed as we drove up to this hangar, the hangar doors are remotely controlled. The OpsCenter controls access to the hangars via an Internet connection to keep unauthorized individuals out of the hangar, and to make it easier for our members to enter the hangar. The OpsCenter uses video cameras mounted outside the hangar to assess whether or not they should grant access."

"This system of observation and controlled access yields several benefits to the club," Arnie said. "For example, the company we lease the aircraft from has lowered our lease rates because they know that their assets are secure in our hangars, and our insurance underwriters have lowered our insurance rates for the same reason. And our members like the extra feeling of security they get when they're in the hangar at night and/or when no one else is around." "What I like best about the system though," Arnie shared, "is the peace of mind it gives me because I know the airplane I'm about to go risk my life in has not been intentionally or unknowingly mishandled or sabotaged."

"Now," Arnie said, "I'd like to turn this briefing over to Jack." "He's going to give you some safety tips and explain what he does here in the hangar to support our members." With that, Jack traded places with Arnie in front of the SR22 and addressed Alpha Flight. "First of all," Jack began, "it's very important for all of us to have good situational awareness whenever we're around airplanes, even when they're parked in a hangar." "Let's start by talking about this propeller," he said as he pointed to the prop on the nose of the SR22. "We have to assume that there might be fault in the aircraft's ignition system that would allow the engine to turnover or start if we move the propeller. So, we always treat the prop as if it could come alive and try to hurt us."

"You will also note that there are many sharp edges and pointy things sticking out of this airplane. Therefore, when walking around it you have to guard against being distracted and bumping into something that might hurt you. With that in mind, let's break up into two teams and Arnie and I will show you how easy it is to get in and out of the airplane."

Half the group followed Jack as he moved around the left wingtip to where the wing met the fuselage of the airplane. The other half followed Arnie to the same point on the right side of the airplane. When everyone was assembled, Jack and Arnie showed each of them how to get safely in and out of both the front and rear seats. They gave everyone enough time sitting in the seats to get a feel for what it would be like to fly in the SR22.

After everybody had sampled the delights of the SR22's cabin, Jack and Arnie led the group back to the front of the airplane. Jack continued his briefing by saying, "My primary job responsibility is to assist our members in getting their aircraft ready to fly, and I help them put it in the hangar and secure it after a flight." "For example, before a member arrives at a hangar for a flight, my crew and I fully prepare the aircraft for the flight by performing a thorough preflight inspection to make sure it's airworthy. We also make sure its looks like new both inside and out. We validate that the proper amount of fuel is on the airplane and we put any catering the member may have ordered for the flight in the airplane."

"If the weather is good, we pull the airplane out and stay with it until the member arrives at the hangar. If the weather is bad, we leave the airplane in the hangar where the pilot does his preflight inspection. Then, while the airplane is still in the hangar, we load the luggage and the pilot and passengers board the airplane. When everything is ready to go, we pull the airplane out of the hangar with the people snug in the cabin."

Chapter 3: Airmanship 2.0 In Practice

"In either case, we load any luggage the member may have onto the airplane and assist the member and his passengers with boarding it. We assist and backup the member in preflighting the airplane and perform 'fire guard' duties as the member starts the engine. We then wave him off as he departs. Of course, if there are any problems, we're right there to help the member resolve them quickly."

"If the member has driven his personal vehicle to the airport, we clean it inside and out and put it in the hangar for the member's return. We will also, if the member requests it, take the car to and from any maintenance facility the member specifies, and we make sure it's back in the hangar for the member's return. Members can also, if they so choose, leave their personal vehicles in the clubhouse's secure parking lot and take the crew van from the clubhouse to and from the hangar."

"Whenever a club airplane returns from a flight, we are here to meet it. We help the member and his passengers to deplane and unload any luggage that may be on the airplane and put it in the member's vehicle. We then complete a post-flight inspection of the airplane and clean the outside and inside so that it once again looks like a brand new airplane for the next flight. When that's done, we secure the airplane in the hangar."

"A member can also choose to have the airplane brought over to the ramp adjacent to the clubhouse where it will be waiting for him under the large canopy you saw. In that case, we perform all of the services I told you about on the ramp in front of the clubhouse." "Now," Jack said, "let me turn things back over to Arnie for the remainder of your hangar tour."

Arnie stepped back in front of Alpha Flight and said, "If you'll follow me, I'll show you a couple of other features here in our hangar complex that our members enjoy." Arnie led the group out of the hangar and around to the end of the hangar building. At the end of the building, Arnie escorted them through a door that led into a well-appointed mini clubhouse. There were comfortable chairs arranged in a way that obviously facilitated conversations and a few small tables equipped with chairs. The group also noted the large-screen TV mounted on one of the walls and a kitchen area with a refrigerator, sink and microwave. There was a sign over a door at the back of the room that read "Airman Briefing Area". Arnie led the group through that door.

"In the Airman Briefing Area," Arnie told them, "we have individual cubicles that are equipped with three chairs, a computer workstation and a large monitor." "We can use these briefing cubicles before and after a flight. If it's an instructional flight, the member and his flight instructor brief the flight beforehand and then debrief it afterwards in one of these cubicles. You'll hear more about that later. If it's a personal flight, the member can use the cubicle

183

to hold an online meeting with a flight dispatcher in the OpsCenter. The flight's dispatch release will be delivered to the member here. Or, the member can do that face-to-face over in the OpsCenter before he comes to the hangar. Whichever is more convenient for the member."

Arnie then led Alpha Flight back into the lounge area and said to them, "Well, that about wraps up our tour of our hangar facilities." "If you have any questions, please make sure you ask them during the Q&A session we'll have after lunch. Now we're going to return to the clubhouse for three more briefings. Before we go though, I'd like to pose the aviation-related question I mentioned before we left the clubhouse. The first one of you to answer it correctly will have the option of riding back over to the clubhouse with me. So, please raise your hand if you know the answer. Here's the question. Who was the first airman to fly solo across the Atlantic Ocean?"

Just about everyone raised his or her hand. Arnie said, "Wow, it looks like that was too easy." "So, I'll add a second part to the question. In what year did he make the flight?" All hands except one were lowered. It was Dan's. Arnie then encouraged Dan to give his answer. "It was Charles Lindbergh and the year was 1927," Dan said. "That's right," Arnie said, "come along with me Dan." Arnie and Dan got into the golf cart while the rest of Alpha Flight joined Jack in the crew van. When everyone was aboard, they drove from the club's hangar complex back to the clubhouse.

As Arnie and Dan followed the crew van, Arnie asked, "Dan, why are you here today?" Dan quickly replied, "I've been around aviation for a long time." "I'm sixty-five now and I learned to fly way back in the '60s. Then I flew helicopters for the Army during the Vietnam War. I dropped out of flying for a while to start a family and build a business. When I decided that I wanted to expand the business into a regional market, I joined a flying club and used its airplanes for business travel. I continued to fly increasingly higher performance airplanes right up to the time I retired." "That's great," Arnie said. "What was the most-complex airplane you flew?"

"Well, as you know, complexity is often in the eye of the beholder. I flew a **Beech King Air B90** for several years. It was certainly more complex than the **Beech Baron** I flew before that. When I retired, I was flying the company's **Beech Premier 1A**. The Premier was a very nice little jet and it certainly had much more performance than the King Air, but in my opinion the Premier is less complex than the King Air."

"I know what you mean," Arnie said. "You're certainly a seasoned aviator. I'd like to spend more time with you swapping flying stories, but here we are at the clubhouse. Please tell me quickly why you're here today." I'm here," Dan replied, "because I'm looking for a personal-flying community that

I can join so that I can continue to enjoy personal flying in a safe and affordable manner." "I'm pretty particular about what I want out of that community. Let me tell you, from what I've seen so far, I think I'll be applying for membership in the Palwaukee Aero Club."

"That's gratifying to hear Dan," Arnie said sincerely, "especially coming from an experienced personal flyer and professional pilot." "Let me know what you're thinking when we wrap up today. I'll be happy to help you find out anything else you may need to know to make an informed decision about applying to join the club. Now, let's catch up with the others."

Briefing 4: Club Training Programs

A few minutes later, Arnie and Dan rendezvoused with the rest of Alpha Flight in the classroom they had been using for the briefings. The members of all five flights took their seats as Diane once again moved to the front of the classroom. Diane called the briefing to order and said, "I hope you enjoyed your tour of our clubhouse and hangar facilities. I can tell you that you will rarely find private aircraft parked in hangars that nice. And the hangar lounges and briefing rooms, not to mention this clubhouse, are amenities most personal flyers only dream of. Now, I'd like to remind you that if you have any questions about anything you've heard or seen so far, please hold them for the Q&A session that will be held after lunch. And with that, let me introduce our next briefer. He's the chair of the club's Training Operations Committee and he is a very experienced professional pilot and long-time flight-and-ground instructor. So, I'll now turn you over to Josh."

Josh walked to the front of the classroom and said, "I'd like to add my welcome to you on behalf of the members of the Palwaukee Aero Club." "I'm going to brief you on the training programs we use in the club to initially qualify our members to fly our aircraft and then to keep them qualified to fly them. Let me start out by saying that all of our airmanship-training programs are developed, maintained and delivered by the Center For Airmanship Excellence. The programs are designed to assist our members in their pursuit of airmanship excellence. The programs fit into an overall curriculum and each course has a formal syllabus. Each lesson in the airmanship-training programs is 'scenario based'. That means instead of just going out and practicing maneuvers like it is done in most general-aviation flight schools, each lesson follows a real-world scenario that is designed to achieve the training objectives for the lesson and also to provide the member with the experience he will need to perform his duties as an airman when the training program is completed."

Josh continued, "All of the Mentor Pilots, flight instructors, simulator instructors and ground instructors who provide flight-training services to our members are certified by, and work for, the Center For Airmanship Excellence. The Center also manages our airmanship-training programs and audits the results to make sure that the programs are being effectively delivered to our members. Our airmanship-training programs are integral components of every member's Personal Airmanship Development Plan. We'll talk more about that later."

"I'd also like to point out that the type airmanship-training programs our members select for inclusion in their Personal Airmanship Development Plan are based on what the member needs to become and remain fully qualified to fly our aircraft. For example, a seasoned airman usually needs to complete only some of the training while someone just starting out on his pursuit of airmanship excellence will go through most, if not all, of the courses. I'm going to brief you on all of the courses even though you may not actually complete every one of them."

"We don't have a whole lot of time to look at our many airmanship-training programs in detail in this briefing, so let's call this a 35,000-foot flyby," Josh said. "This briefing will only provide you with the essentials. However, as you'll see, the first airmanship-training program you'll go through with us will provide you with a very detailed look at all of our training programs. We use six airmanship-training programs that were, as I mentioned, developed by the Center For Airmanship Excellence," Josh told them. "They are the 'Introduction To Personal Flying Program', the 'Sport Pilot Qualification Program', the 'Personal Mobility Pilot Qualification Program', the 'Recurrent Training Program', the 'Professional Pilot Preparation Program' and the 'Airmanship Challenge Program'."

"The Introduction to Personal Flying Program consists of two courses. They are the Discover Personal Flying Course and the Solo Pilot Course. The Discover Personal Flying Course is designed to provide you with the knowledge and experiences you'll need to decide if learning to fly is for you. The course begins with a two-hour classroom session that is an overview of everything you'll have to know and do to become fully qualified to fly the club's aircraft. You'll also be assigned some directed self-study. That's followed by a session in a **Cub simulator** with a flight instructor. Next, you will experience a flight lesson in a real Cub. Then, you'll have a training session in a CTLSi simulator followed by a flight lesson in a CTLSi. After that, you'll experience a simulator-training session in an SR22 simulator with a flight instructor. Then, you'll move on to a flight lesson in an SR22. This lesson is designed to give you a feel for how the SR22 flies and to provide you with a real personal-mobility experience so that you can find out what it's like to travel by private aircraft as the Pilot In Command."

"Let me point out," Josh emphasized, "every simulator and flight-training session in all of our airmanship-training programs includes a preflight briefing and a post-flight debriefing with an instructor. Some directed self-study is also assigned for every training session. We record every simulator- and flight-training session. The recording includes video from five cameras mounted in various locations on and in the airplane, audio of intercom and radio conversations and flight data like airspeed, heading, altitude and position. The member and his instructor use these recordings in their pre-flight and post-flight debriefings. The member can also review the recordings anytime, and as many times, as he wants via an Internet connection. This is a tremendously effective training tool that reduces training time and expense and adds remarkable value to our flight-training experiences. This important training tool is not offered by most general-aviation flight schools."

"Upon completion of the Discover Personal Flying Course, you will have a very good idea as to whether or not you want to learn to fly and what it's going to take to do that. You will also have had some really interesting and exciting personal-flying experiences. Most of our members complete the course in less than a week of part-time effort."

Josh continued, "The next course you will take is the Solo Pilot Course." "Just as the name implies, at the completion of this course you will make your first solo flight. All of us who have learned to fly mark our first solo flight as the point at which we formally entered the flying fraternity. You'll always remember it. I still do."

"You will fly the Cub during the Solo Pilot Course. The Cub is what is known as a 'conventional-gear' airplane. We also refer to it as a **Taildragger**. That means that instead of having a nosewheel located near the front like most personal airplanes—we call these 'tricycle-gear' airplanes—it has a tailwheel located below the rudder at the rear end of the airplane," Josh said.

"For primary and advanced flight training, almost all other general-aviation flight schools use tricycle-gear aircraft. The Center For Airmanship Excellence has specified the use of the Cub for primary training precisely because it is a taildragger. The Center did this for several reasons."

"The first reason is that approximately one-half of all personal-flying accidents take place during takeoff, approach to landing, landing and low-altitude maneuvering flight. The root cause of virtually all of these accidents is the pilot's lack of proficiency in basic 'stick-and-rudder' flying skills. These skills need to be solidly acquired in initial training and then maintained over time through recent experience and recurrent training. We believe that the best way to develop and maintain your 'stick-and-rudder' skills is to fly a taildragger."

Josh continued, "That's because a taildragger is much harder to handle on the ground, especially during takeoff and landing, than a tricycle-gear airplane." "And that's because the taildragger's center of gravity is located behind the main wheels instead of in front of them as is the case with a tricycle-gear airplane."

"In a taildragger, if the centerline of the airplane is not perfectly aligned with the direction the airplane is moving when it's in contact with the ground, the tail of the airplane wants to swing rapidly in the direction the center of gravity is moving in. For example, if as you touch down during the landing maneuver the tail of your airplane is cocked off to the left, the tail will want to swing further to the left as soon as the main wheels touch the runway. The opposite is the case with a tricycle-gear airplane. A tricycle-gear airplane has a tendency to straighten itself out after touchdown."

"This 'self-aligning' tendency in tricycle-gear aircraft is usually a good thing. However, it instills poor handling skills in new pilots. If these pilots then continue to fly only tricycle-gear aircraft, their already subpar aircraft-handling skills deteriorate. This is a recipe for loss of control as evidenced by the number of accidents that result from poor pilot technique during takeoffs and landings. Experienced aviators know that learning to fly in a taildragger and then taking frequent recurrent training in one goes a long way in assuring that one avoids becoming an accident statistic."

"Also," Josh said, "we use taildraggers in our primary flight training program because they teach a pilot to make better landings, no matter what type of aircraft he's flying." "For example, when landing a tricycle-gear airplane it is very important to touch down on only the main wheels first and then gently lower the nosewheel to the runway. Pilots who learn to fly in tricycle-gear aircraft have a tendency to 'land flat'. That means they land on all three wheels at the same time. They also tend to let the nosewheel fall to the runway too quickly after the main wheels touchdown. Pilots who learn to fly in a taildragger don't have either of these tendencies when landing tricycle-gear airplanes."

"Another reason to do your initial pilot training in a simple taildragger like the Cub is the growing number of accidents that are caused by pilots who rely too heavily on aircraft instrumentation. Instead of concentrating on the flight instruments, they should be feeling and hearing what the airplane is doing and looking out the window for visual cues that tell him how the airplane is performing. By eliminating all of the advanced flight instruments in the Cub, we can teach these 'stick-and-rudder' skills to new airman more effectively. Our Cubs are equipped with only very basic flight instruments, a simple

Chapter 3: Airmanship 2.0 In Practice

communications radio, a '**Transponder**' that makes it easier for Air Traffic Control to see the airplane on radar and a traffic-alerting system."

"Situational awareness is very important to a pilot's longevity," Josh said. "Learning to fly in the Cub establishes the foundation for good situational awareness. This is because there is no radio- or GPS-navigation equipment installed in the airplane. This requires new airmen to learn and perform the very basics of aerial navigation. It has been shown over and over again that skipping this crucial step in a pilot's initial flight training makes it considerably more difficult for him to maintain situational awareness, even in a glass-cockpit aircraft."

Josh went on, "Another very good reason to learn how to fly in the Cub is that it is just pure fun." "Since the 1930s, tens of thousands of the personal-flyers, military aviators and professional pilots in the U.S. have taken their primary training in an original Piper Cub. Virtually all of them sing its praises as a primary flight trainer and everyone who has ever flown in a Cub will tell you that it really is fun to fly. There's nothing like strapping into a Cub on a warm summer evening, fully opening the door on the right side and the window on the left and starting the engine for a few trips around the traffic pattern as dusk falls. There are few better ways to experience the sights, sounds and smells of flight. It really makes you feel like an early aviation pioneer who is in touch with the essence of flight."

"You'll practice your takeoffs and landing in the Cub, and solo it for the first time, on a grass strip, or runway. Just as the name implies, the runways you'll use for this phase of your training are not paved, but are surfaced with grass. This makes it a little easier to maintain directional control of the Cub during takeoffs and landings, and it makes the experience more enjoyable."

"So," Josh summed up, "that's why we use the Cub as the starter airplane for all of our new members." "However, let me be clear that we don't fly antique Cubs. The Cubs we fly were built within the past five years. The old Cubs don't have electrical systems. Pilots have to spin the propeller by hand to start the engine. You can imagine how dangerous that is. Our Cubs have a modern electrical system that supplies electrical power for the starter, lights and the avionics equipment in the airplane. Our Cubs also have modern safety devices like air bags and a whole-aircraft-parachute system. So, I think you can now see why we have Cubs in our club fleet. It's a pity that our club is the only organization in the Chicago area that provides primary flight training in a taildragger of any kind, let alone the iconic Cub."

Josh directed the briefing back on course by saying, "As I mentioned before we went into the details on the Cub, you'll be flying the Cub in the Solo Pilot Course." "The Solo Pilot Course consists of forty-two hours of directed self-study, ten hours of in-classroom ground school, ten hours of Cub simulator training, ten hours of Cub flight training and a one-hour preflight briefing and a one-half-hour post-flight debriefing with each flight-training session. The course is capped with a one-half-hour solo flight in the Cub. 'Solo' means you'll be flying the Cub all by yourself."

"The in-cockpit video recording of your first solo, plus an external video shot by your instructor of your first solo landing, will be provided to you to commemorate this important milestone in your pursuit of airmanship excellence. Most of our members complete this course in about five weeks of part-time effort," Josh said. "When you complete it, only approximately six weeks after you began your training, you will feel like a real aviator and you will be fully prepared for the next program in the curriculum."

"That next program on your path to airmanship excellence is the Sport Pilot Qualification Program," Josh told them. "There are four courses in this program. They are the 'Sport Pilot Certification Course', the 'Advanced Aircraft Attitude Management Course', the 'CTLSi Transition Course' and the 'VMC Into IMC Management Course'. When you complete this program, you will be fully qualified to fly the club's Light Sport Aircraft—LSAs—like the CTLSi, during the day and in good weather. You'll be able to take one passenger with you on your flights to anywhere in the U.S. and you will be certificated by the FAA as a Sport Pilot."

"During the Sport Pilot Certification Course, you will spend ten hours in the Cub simulator and thirteen hours flying the Cub," Josh told the group. "Eight of the flight hours in the Cub will be with an instructor, we call that 'dual flight training', and five hours will be solo. You will be assigned a little over seventy hours of directed self-study, attend a ten-hour ground school and you will receive preparation for the FAA Private Pilot Knowledge and Practical tests and the Sport Pilot checkride."

"After you have completed the directed self-study and the ground school, you will take an FAA Knowledge Test. And at the end of the Sport Pilot Certification Course, you will take an FAA checkride and you will be issued your FAA Sport Pilot Certificate. It will take you approximately two months to complete this course with a part-time effort."

"The next course you will take in the Sport Pilot Qualification Program will be the Advanced Aircraft Attitude Management Course. This course is essential for safe personal flying. You won't find this course as required training in any other general-aviation flight school, flying club or aircraft-rental operation. Yet, approximately six percent of all personal-flying accidents are attributed to loss of aircraft control during maneuvering flight close to the ground."

"This loss of control results from the aircraft exceeding its performance limitations. When this happens, an accident is inevitable if the pilot doesn't know how to regain control of the airplane. You will learn how to recover from a loss of control in the event that you should experience one. The course consists of ten hours of directed self-study and a two-hour ground school. This is followed by three one-hour flight-training sessions in an aerobatic airplane. It will take you approximately two weeks of part-time effort to complete this course."

"Your next milestone in the Sport Pilot Qualification Program is the CTLSi Transition Course," Josh told them. "In this course, you'll get comprehensively checked out in the Flight Design CTLSi. The course consists of eighteen hours of directed self-study and a four-hour ground school. You will also train in the CTLSi simulator for one-and-one-half hours and fly the CTLSi with an instructor for six hours. This course should take you about two weeks to complete on a part-time basis."

"And the final course in the Sport Pilot Qualification Program is the VMC Into IMC Management Course. 'VMC' stands for 'Visual Meteorological Conditions' and 'IMC' stands for 'Instrument Meteorological Conditions'. When you're flying in VMC, you fly the airplane by reference to what you can see outside the airplane—the sky and the ground. In IMC, you fly the airplane by reference to the flight instruments in the airplane," Josh said.

"VMC-into-IMC related accidents occur when a non-instrument-qualified pilot inadvertently flies his aircraft into clouds or fog. A pilot can also lose control of his aircraft under certain conditions at night if he's not capable of flying his aircraft by reference to his flight instruments. These types of accidents account for approximately half of all weather-related personal-flying accidents. Over eighty-five percent of them are fatal. In fact, when a pilot who is not capable of flying his aircraft by reference to his instruments enters a cloud, he has approximately 178 seconds remaining in his lifespan."

Josh went on to say, "The Palwaukee Aero Club's values of shared-decision-making and checks-and-balances do a great job of keeping its members out of situations that may lead to a VMC-into-IMC accident, but we believe that it is still possible for one of our members to get into a situation where the ability to fly his airplane by reference to his flight instruments is crucial." "The VMC Into IMC Management Course provides our members with the skills, strategies and tactics they will need if they find themselves inadvertently in the clouds."

"The course consists of a two-hour ground school, three hours of directed self-study, a one-and-one-half-hour session in the CTLSi simulator with an instructor, ten hours in the CTLSi simulator by yourself and a one-and-one-half-hour training flight in the CTLSi with an instructor. This course should take you about two weeks to complete."

"I'd like to point out," Josh said, "that to my knowledge we are the only flight organization that offers this type of course or requires the pilots who fly in the organization to demonstrate their ability to extricate themselves from a VMC-into-IMC situation. In my opinion, any pilot who flies at night or on long cross-country flights should have this capability."

"When you complete the Sport Pilot Qualification Course, you will be fully qualified to fly the club's Light Sport Aircraft like the CTLSi. You can reach this point in less than five months of part-time effort from the time you start your airmanship training. You can decide to stay at the Sport Pilot level and fly only the club's Light Sport Aircraft during the day. As you'll recall, you will be allowed to take only one passenger with you as a Sport Pilot."

"If you decide that you want to become fully qualified to fly the club's Cirrus SR20s and SR22s, you'll move on to the Personal Mobility Pilot Qualification Program. While you're going through this program, you'll be able to exercise the privileges of your Sport Pilot Certificate in the CTLSi for your personal flying. Sport Pilots are prohibited from personally flying for business-related reasons. Therefore, if you want to maximize your personal-mobility options, you'll want to complete the Personal Mobility Pilot Qualification Program."

"When you complete the Personal Mobility Pilot Qualification Program," Josh began, "you will be fully qualified to fly the club's Personal Mobility Aircraft like the Cirrus SR20 and SR22." "The program is made up of five courses—the Private Pilot Certification Course, the Instrument Rating Preparation Course, the SR20 Transition Course, the Instrument Rating Certification Course and the Cirrus SR22 Transition Course."

"In the Private Pilot Certification Course, you'll continue to fly the Flight Design CTLSi. The course consists of thirty-five hours of directed self-study, a ten-hour ground school, twelve CTLSi-simulator sessions, seven CTLSi dual-flight-training sessions, five CTLSi solo-flight-training sessions, preparation for the FAA Private Pilot Knowledge and Practical tests and a checkride. When you complete this course, you will have earned your FAA Private Pilot Certificate. This is another important milestone on your way to becoming a fully qualified airman. At this point, there will be only three more steps to complete."

"And the next step is," Josh said, "the Instrument Rating Preparation Course." "Let me give you a little background on this course. It is perfectly legal for a brand new Private Pilot to go gallivanting off across the landscape with as many passengers as his airplane can hold. Unfortunately, this is not a very good idea for several reasons, but the most important reason is that sooner or later that new pilot is going to encounter Instrument Meteorological Conditions, dark nights, snow showers or other aeronautical dragons that will challenge him beyond his capabilities to control an airplane solely by reference to the flight instruments. As I mentioned earlier, if this happens, things get real ugly real fast."

"For this reason, as well as a few others, the Center For Airmanship Excellence specifies that a fully qualified Cirrus SR22 pilot hold an FAA-issued instrument rating. This rating allows a pilot to legally fly his aircraft in Instrument Meteorological Conditions. As you know, the Palwaukee Aero Club adheres to the Center's specifications because we believe that the Airmanship 2.0 principles underlying those specifications will keep our members safe. Also, an instrument rated pilot tends to be a more-efficient and self-confident airman."

"The purpose of the Instrument Rating Preparation Course is to prepare you for the Instrument Rating Certification Course. To qualify for an Instrument Rating, an airman has to meet certain total flight time and cross-country flight time requirements. Those requirements are met in the Instrument Rating Preparation Course. The course consists of eighty hours of directed self-study, five CTLSi-simulator sessions that focus on instrument flying and twenty-five cross-country flights under visual and simulated-instrument conditions. At the end of this course, you will be ready to enter the Instrument Rating Certification Course," Josh told them.

"You'll be flying the SR20 in the Instrument Certification Course, so before you start that course, you'll take the Cirrus SR20 Transition Course. This course consists of sixteen hours of directed self-study, a four-hour ground school, four Cirrus SR20-simulator sessions and four SR20 flight lessons," Josh said. "When you complete this course, you will be ready to fly the SR20 in the Instrument Rating Certification Course. However, if you also want to fly the club's SR20s for your business-and-personal-mobility needs you'll have to complete the Instrument Rating Certification Course first."

"The Instrument Rating Certification Course prepares you to become a safe and proficient instrument pilot. It consists of a little over one hundred hours of directed self-study, a twelve-hour ground school, twenty Cirrus SR20-simulator sessions, ten SR20 flight lessons, preparation for the FAA Instrument Rating Knowledge and Practical tests and a checkride. At the completion of this course, you will be fully qualified and legal to fly an SR20 in Instrument Meteorological Conditions."

"The last course in the Personal Mobility Pilot Qualification Program is the Cirrus SR22 Transition Course. This course consists of sixteen hours of directed self-study, a four-hour ground school, four, Cirrus SR22-simulator sessions and four SR22 flight lessons," Josh said. "When you complete this course, you will be fully qualified to fly the club's SR20s and SR22s in the daytime and night time, and in Visual Meteorological Conditions and Instrument Meteorological Conditions. You will reach this point on your quest for airmanship excellence within about twenty-eight weeks from the time you started the Personal Mobility Pilot Qualification Program, and in a little less than one year from the time you began your airmanship training with the Discover Personal Flying Course. This timeline projection is based on a part-time effort. If you can devote more time to the program, you can complete the training in much less calendar time."

"When you complete your initial-qualification program, whether it be only the Sport Pilot Qualification Program or the Personal Mobility Pilot Qualification Program too, you will still have a lot of training ahead of you. In fact, it's a lifetime of training. This training includes periodic recurrent training, transition training into different types of aircraft, upgrade training to increase your airmanship capabilities and directed self-training."

"As an airman who has recently qualified to fly the club's aircraft, you will be required to take recurrent training every three months. This recurrent training ensures that your airmanship knowledge and skills are kept at the level required for your qualifications," Josh said. "The recurrent training includes directed self-study, ground school sessions, simulator sessions and dual-instruction flight lessons. Your recurrent-training program is built into your Personal Airmanship Development Plan. As your airmanship experience increases, the frequency of your recurrent training will decrease. Our most-experienced airmen undergo recurrent training every six months."

"Every time you want to get qualified to fly a different type of aircraft, you will go through a formal transition program. We sometimes refer to these as 'aircraft-checkout' programs. For example, as you will recall, when you move from flying the Cub to the CTLSi, the CTLSi to the Cirrus SR20 and from the SR20 to the SR22 in the training programs we talked about earlier, you will go through a formal aircraft-transition program. This program consists of directed self-study, ground school sessions, simulator-training sessions and dual-instruction flight lessons."

Josh continued, "As an airman who practices Airmanship 2.0, you will be continuously working to increase your airmanship capabilities." "For example, let's say you want to fly the club's twin-engine airplanes. To fly them legally, you have to add an FAA Multi-Engine Rating to your Private Pilot Certificate. To do that, you will go through an upgrade course that will prepare you for an FAA checkride and provide you with the knowledge and skills you will need to

Chapter 3: Airmanship 2.0 In Practice

fly the airplane safely and efficiently. The upgrade course will consist of directed self-study, ground school sessions, simulator-training sessions and dual-instruction flight lessons."

"In addition to the periodic training that I just described, you will also be undergoing continuous self-training. Your Personal Airmanship Development Plan will include training that you can do when you are flying without an instructor. For example, your Personal Airmanship Development Plan might include learning an additional capability of your computerized flight-guidance system that you didn't have to learn to achieve a basic-qualification level. Your plan might include the requirement that on the next five flights that you take, you will practice the use of this additional capability. Of course, the plan will also include some directed self-study and some time in a simulator where you can practice the use of this additional capability before you try it in the airplane."

"Now, I'd like to point out the fact that there are ten young people here today with their parents," Josh said. A rustle went through the classroom as everyone looked around and smiled at the young folks in the group. They all looked to be of high school age.

When things settled down, Josh continued, "All of these budding airmen and airwomen are members of the National Association of High School Aviation Clubs. As I'm sure you've noticed already, we use a lot of acronyms in aviation. So, we call this organization N-A-H-S-A-C or 'nay-sack'. The Palwaukee Aero Club, the Center For Airmanship Excellence, Lewis University, Southern Illinois University, Kishwaukee College and the Chicago chapter of Women In Aviation International all co-sponsor NAHSAC. All of these students who are with us today attend area high schools that have aviation clubs in their schools that are part of the NAHSAC network."

"So, why do does the Palwaukee Aero Club co-sponsor NAHSAC?" Josh asked. "We do it because the club needs a constant flow of entry-level people to staff the club's ground crew. And we do it because part of the Airmanship 2.0 philosophy we subscribe to is spreading the word about the pursuit of airmanship excellence. What better way is there to set an aspiring pilot on the road to airmanship excellence than to expose him or her at an early age to the day-to-day reality of that pursuit as it is manifested by the members of our club?"

"We also co-sponsor NAHSAC because we prefer to 'home grow' our cadre of professional pilots who support our club as instructors, Mentor Pilots, charter pilots and utility pilots. We refer to these airmen as 'Flight Pros'. These young aspiring aviators here with us today have the opportunity to join our club as Student Members and go through the Center For Airmanship

Excellence Professional Pilot Preparation Program. A stint as a Flight Pro is part of that program."

"And we support NAHSAC because every member of the Palwaukee Aero Club is dedicated to expanding the ranks of airmen who practice Airmanship 2.0. The NAHSAC high-school-aviation-club network in our area funnels not only high-school age Student Members to our club, but also regular adult members who hear about the Palwaukee Aero Club from the students. These young folks are here with us today because they are very interested in finding out about how they can qualify for one of the very limited slots we offer in our Professional Pilot Preparation Program."

"Their parents are here with them because they realize that they will have to support their son or daughter in achieving their dream of becoming a professional airman, and they know that they need a comprehensive and honest briefing on the best way for their child to achieve that dream. Our hope is that by the end of today, these students and their parents will be able to make a highly informed decision on whether or not to enroll in our Professional Pilot Preparation Program."

"They will receive a separate, more-detailed briefing on that program later today, but I'd like to take just a couple of minutes to describe the program to you in case you have children, or you know someone, who might be interested in pursuing a career as a professional aviator," Josh said. "The Professional Pilot Preparation Program, just like all of the training programs we use in the Palwaukee Aero Club, was conceived and developed by the Center For Airmanship Excellence. The Center also delivers, maintains and administers the program. The folks at the Center who decide what should go into an excellent career-track program are some of the most experienced professional aviators on the planet. Many of them also have over half a century of professional experience in the flight-training industry. And they come from airline, military, corporate, government-flying, general-aviation-flying and aviation-management backgrounds."

"As a professional pilot myself," Josh said, "I can tell you that I wish this program had been around when I was in high school." "Nothing like it existed then and it is still a rarity today. We feel very fortunate to be able to offer this opportunity to young people from our community who are committed to, and capable of, making their dream of professional flying a reality. The core design of the Professional Pilot Preparation Program is focused on providing someone who truly desires to become a professional pilot with the guidance, encouragement, assistance and mentoring that it takes to make him or her that one-in-a-hundred successful candidate for a professional-flying job. A big part of the guidance we provide is sharing with aspiring pilots and their parents the

Chapter 3: Airmanship 2.0 In Practice

three entry tracks that have been proven to lead to professional-piloting job offers. These tracks sometimes cross over or even intertwine for a while. But virtually every professional pilot in the U.S has walked one, or a combination, of these paths.

The entry portal to the first of these three paths is located in the admissions office of a highly respected aviation university like Embry-Riddle Aeronautical University, Lewis University or Southern Illinois University. The second portal is the door of a U.S. Air Force, U.S. Army, U.S. Navy, U.S. Marine or U.S. Coast Guard recruiting office. The third portal is located in general-aviation flight schools."

"Deciding which one of these portals to enter first is dependent upon the resources you can invest in your quest to become a professional pilot and the particular flavor of professional flying you'd prefer for a career—military, airline, corporate, charter, general aviation, etc. The Center helps aspiring pilots and their parents to make the right choice for their desires and situation. This avoids the squandered resources and dashed hopes that result from going through the wrong portal first."

"The most important thing to keep in mind no matter which path you choose," Josh went on, "is to arrive at your chosen portal as well qualified as you can and as soon as you can." "This approach can significantly reduce your costs in preparing for, and appreciably increase your chances of reaching your end goal of, flying professionally. And the earlier you arrive at your launch portal, the fewer people are ahead of you and the more people are behind you on that path."

"In most cases, a professional pilot's 'seniority', how long he has been flying for a company in relation to how long every other pilot has been with the organization, will determine how much he earns over his career, where he lives, what types of aircraft he flies, his days off, his vacation time and a score of other lifestyle issues. Even if seniority isn't necessarily an issue, for example as in corporate flying, the earlier a pilot starts building his professional expertise the more 'seniority' he has when it comes to the same lifestyle issues."

"So, to provide you with a better picture of how all this works, I'd like to give you a quick overview of a typical career track that would be recommended by the Center For Airmanship Excellence for a young man, let's call him 'Joel', who is currently fourteen years old and in the fall of his freshman year in high school. We're also going to assume that Joel is a member of the NAHSAC-affiliated aviation club at his high school and that Joel and his parents have attended a briefing like this one."

"After that briefing, Joel and his parents met with a Center For Airmanship Excellence Mentor Pilot to make sure they fully understood their options and what is required to pursue those options. This is the type of meeting the students here with us today and their parents are going to attend later this afternoon. In this hypothetical but typical case, Joel and his parents have clearly focused on a career as an airline pilot for Joel. After deliberating and getting all their questions and concerns addressed, they decided to put Joel on the most-efficient and shortest path to that goal."

"Joel, his parents and a Mentor Pilot put together an airmanship-development plan that starts out with Joel continuing to participate in his high-school aviation club during the school year and following a regimen of directed self-study of aviation topics. In the summer of his fifteenth year, which would be following his freshman year, Joel will attend an Airmanship 2.0 summer camp wherein he'll receive ground and flight training that will prepare him for a solo flight in a glider. The summer camp will also immerse him in an Airmanship 2.0 culture. At the end of the camp, Joel will fly a glider by himself for the first time. We refer to this as 'soloing a glider'."

"When Joel returns to school in the fall for his sophomore year, he will continue his participation in the aviation club and he'll follow a plan of directed self-study of airmanship topics. He'll also enroll in the Palwaukee Aero Club Introduction To Personal Flying Program. He will begin his Discover Flying Course shortly before Thanksgiving and complete it by the end of the calendar year. Joel will then enroll in our Solo Pilot Course while he continues to participate in his high school's aviation club. He will complete the Solo Pilot Course when he turns sixteen with his first solo flight in a powered airplane."

"During the summer following his sophomore year, he will attend another Airmanship 2.0 summer camp where he'll begin the Sport Pilot Qualification Program. He'll work his way through that program over the course of his junior year while continuing to attend aviation-club meetings. He'll complete the Sport Pilot Qualification Course and receive his FAA Sport Pilot Certificate when he turns seventeen during his junior year in high school."

"During the summer following his junior year when he is seventeen, Joel will attend another Airmanship 2.0 summer camp where he'll enter the Personal Mobility Pilot Qualification Program. When he returns to school in the fall as a senior, Joel will continue to attend aviation club meetings and progress through the Personal Mobility Pilot Qualification Program."

"He'll earn his FAA Private Pilot Certificate before Christmas of that year and he will complete the Personal Mobility Aircraft Pilot Qualification Program by the time he turns eighteen. When he graduates from high school, he will have earned his FAA Sport Pilot Certificate, FAA Private Pilot Certificate and his FAA Instrument Rating. And he will have been steeped in an Airmanship 2.0 culture for about four years. Certainly more than enough time to instill in an ambitious young man or woman the values and beliefs he or she needs to be a success in the world of professional flying."

"Following his high-school graduation, Joel will enter our full-time Professional Pilot Preparation Program. This program has been designed to comprehensively prepare him for an entry-level position as a professional pilot. The program will fully qualify Joel for his FAA Certified Flight Instructor Certificates (there are three), his FAA Commercial Pilot Certificate and his FAA Multi-Engine Rating."

"From the time Joel enters the Professional Pilot Preparation Program until he is qualified to go to work as an entry-level professional pilot, he will work for the Palwaukee Aero Club as a paid intern. In this roll he will serve on the club's ground crew and perform other airmanship-related duties. When Joel is fully qualified for the entry-level professional-pilot position, he will be hired by the Center For Airmanship Excellence at a competitive rate of compensation and assigned to support one or more of the aero clubs and other flight organizations that the Center supports," Josh told them.

"Joel will work as a flight instructor, ferry pilot and all-around utility pilot for about two years while he is eighteen and nineteen years old. During this time, he will build his total flight time to around 750 hours, continue a regimen of directed self-study and ground-school sessions related to airmanship development, complete regular assigned training courses and earn enough college credits to qualify for an associate's degree from Kishwaukee College's Aviation Flight Program."

"During Joel's twentieth and twenty-first years, he will continue to fly as a professional pilot in the capacities I mentioned, and in addition he will fly as a copilot on business jets. During this period, he will build his total flight time to around 1,500 hours, continue a regimen of directed self-study and ground-school sessions related to airmanship development, complete regular assigned training courses and earn enough college credits to qualify for a Bachelor of Science in Aeronautics degree from Embry-Riddle Aeronautical University through its Embry-Riddle Worldwide program."

"During Joel's twenty-second year, he will continue to fly as a professional pilot in the capacities I mentioned. He will build his total flight time to around 2,000 hours, continue a regimen of directed self-study and ground-school sessions related to airmanship development and complete a training course that will culminate in him earning is FAA Airline Transport Pilot Certificate at the age of twenty three. This is the minimum age at which the FAA will issue this certificate to him."

"At this point, Joel's qualifications will put him at the head of the line in the highly competitive airline-pilot marketplace. In fact, he may be so well qualified that he'll be able to skip over the regional-airline stepping stone that most other aspiring airline pilots have to serve time on. With any luck, he will be hired directly by a major airline at the age of twenty-three."

"His competitiveness will be greatly enhanced by the fact that all through his preparation for a career as an airline pilot, well-respected professional airmen have mentored him. These mentors, along with the Center For Airmanship Excellence, will highly recommend him to the airlines that are considering hiring him. Also, the quality and comprehensiveness of the Airmanship 2.0 training he has received will give him another competitive edge in the marketplace. The airlines highly value this approach to preparing a pilot for their cockpits. It reduces their upfront training costs and ensures that a new-hire pilot like Joel will be successful in their Airmanship 2.0 culture."

"It will take Joel a total of five years after graduating from high school to reach his career goal of becoming a pilot for a major airline. Other aspiring airline pilots in his age group will typically put in a lot of toil and sacrifice over at least twice as many years to reach the same goal if they follow a more-traditional general-aviation path. Joel, and more importantly his parents, will have spent less than half of what the parents of these other young people spent to get their child to their career goal."

"But more importantly, as a professional pilot in his first five years after high school, Joel will be self-supporting and earn more than what it costs his parents to pay for his training. The parents of his peers will still be supporting their child during these five years and in some cases for several years thereafter. But even more important than this is the fact that Joel got hired by a major airline while ninety-nine out of a hundred of his peers did not."

"So, by enrolling in the Professional Pilot Preparation Program while still a freshman in high school, Joel will be able to reach his career goal about five years ahead of his contemporaries. This translates directly into a tremendous seniority advantage that can increase Joel's career earnings by a few million dollars and provide him with other lifestyle benefits that would take him a lot longer to acquire if he had not selected the Professional Pilot Preparation Program as his path into professional flying."

Josh continued, "I'd like to tell you about one other Center For Airmanship Excellence airmanship-training program that all of our members are continuously enrolled in. That's the Airmanship Challenge Program. The Airmanship Challenge Program has two purposes. The first is to continually improve the airmanship of the airmen who participate in it, and the second is to

Chapter 3: Airmanship 2.0 In Practice

have fun. We believe that Airmanship Challenges plus flying fun plus recognition plus rewards equals continuous airmanship development."

"Airmanship Challenges are designed to provide an interesting and stimulating challenge that can be accomplished with a bite-sized investment of time, money and effort. Airmanship Challenges are also designed to fit easily into an aviator's airmanship-development plan. Incorporating appropriate Airmanship Challenges into a Personal Airmanship Develop Plan facilitates continuous airmanship development and leads to the attainment of an aviator's airmanship goals."

"For example, the 'VMC Into IMC Airmanship Challenge' tests a non-instrument-rated airman's ability to safely extricate himself from an inadvertent encounter with instrument meteorological conditions. And the 'High Density Altitude Operations' Airmanship Challenge pits an airman's knowledge of high-density-altitude operations and his aircraft-handling skills at high-density altitudes against simulated and actual flight scenarios that include high-density-altitude operations. The Center has developed over one hundred Airmanship Challenges that our members, with the help of their Mentor Pilot, can select from for inclusion in their Personal Airmanship Develop Plan."

"Well, I'm just about done with my briefing on how we go about airmanship training in the Palwaukee Aero Club," Josh said. "But before I wrap things up, I want to address another topic that is more important than anything I've told you so far. That subject is your approach to learning to fly. If you take the wrong approach, you probably will never learn to do it."

"I know that's a shocking statement, but there is data to back it up. The facts are that of every one hundred people who start out to learn to fly at traditional general-aviation flight schools, over eighty of them drop out before they reach their initial pilot-certification goal. That means they never get the FAA pilot certificate that they start out to get. And these folks usually spend well over ten thousand dollars before they finally give up and drop out. Sadly, these poor disillusioned souls almost always abandon their dream of learning to fly for the rest of their lives."

"And most, if not all, of the roughly twenty tenacious student pilots who do finally make it through an initial training program at a traditional general-aviation flight school tell us that there were many, many things wrong with their flight-training experience that almost caused them to drop out several times. Even worse, many of these newly minted pilots either walk away from personal flying after they have achieved their goal of 'learning to fly', or they curtail their flying to the point that they can no longer maintain the precious proficiency that is necessary for survival in the highly dynamic flight environment."

201

"The bottom line is, in my opinion," Josh interjected, "just what General Chuck Yeager said, 'if you're going to fly, do it right'." "That obviously carries over into learning to fly. If you're going to learn to fly, learn to fly the right way."

"Let's do a little thought experiment to put this into perspective. Imagine that you've never learned to drive. You've always felt it was too difficult and dangerous, but now you've decided to learn to drive for personal-mobility reasons. You get a ride to a local strip mall and walk into AAA Driving School with the idea of signing up for professional driving lessons that will prepare you for your state's driver's test. As you enter the lobby you note that all of the people within eyesight are preoccupied with something else and you're ignored. You suddenly feel like an outsider and an unwanted intruder."

"A teenage-looking young man finally notices you when he looks up from his iPhone and asks you if he can help you," Josh continued. "He introduces himself to you as a driving instructor for the school. When you tell him you're thinking about taking driving lessons, he immediately tries to sell you lessons. He suggests that you accompany him out to the parking lot to look at a car the school uses in its course."

"When you first see the car, you notice that it's at least thirty-years old and not in very good condition. Your prospective driving instructor tells you that it's easy to get a driver's license, but he doesn't tell you about the eighty percent drop-out rate the school experiences. Would you hire this driving school to teach you to drive? If you did, you'd be jeopardizing your safety, the safety of the passengers in your car and the other folks on the road with you. Unfortunately, this is typically what happens when a prospective pilot walks into a typical general-aviation flight school. The club approaches our airmanship training more like a Mario Andretti high-performance-driving school than the one I just described."

"I don't have the time in this briefing to go into any more detail about how traditional general-aviation flight schools are not 'doing it right'," Josh said. "There are many more reasons I could share with you if we had the time now. If you want to know more about this subject, there's a lot of information on the club's and the Center For Airmanship Excellence's Websites. Also, when you are enrolled in the Discover Personal Flying course, you will have a homework assignment that will require you to visit other flight schools. You'll be provided with a checklist that will help you to determine if these flight schools are 'doing it right'."

"For now, just let me say that all of us in the Palwaukee Aero Club believe the Center For Airmanship Excellence has designed training programs that are aimed at getting you fully qualified to 'fly right' and not just so that you can pass an FAA checkride to minimum government standards. Every member of our club has been through several of the Center's programs and we can testify that they work as advertised."

"We have an almost zero dropout rate in our Sport Pilot Qualification Program and our Personal Mobility Aircraft Pilot Program. We do have some dropouts in the Introduction To Personal Flying Program because this program is designed to give people the opportunity to find out if personal flying is really for them. Obviously, some find that it isn't. But our experience shows that once a person completes the Introduction To Personal Flying Program and makes the decision to learn to fly using our training programs, that person goes on to become a fully qualified personal flyer."

"Another thing that I'd like to share with you that relates to learning to fly right, is the fact that you only learn to fly once. I don't know of anyone who goes through the time, effort and expense of learning to fly a second time. So, my sincere advice to you is if you're going to learn to fly, do it once and do it right if you want to be a safe pilot. If you learn to fly the wrong way, you run the risk of always being less safe than you could be."

"Now that I've said that," Josh interposed, "I have to clarify it by telling you that people who learn to fly in a traditional general-aviation flight school do learn to fly twice — in a way. They usually end up taking about twice as much flight training than what is required by the FAA. And, they on average take more than three times as long as necessary to achieve FAA certification. They also spend over twice as much money to earn a Private Pilot Certificate as a person who reaches that same level in the Center For Airmanship Excellence Personal Mobility Pilot Qualification Program."

"And, I want to mention another very important aspect of your approach to learning to fly," Josh said. "That aspect is the intensity with which you approach it. As we all know, what we get out of something is directly related to what we put into it. This is definitely the case with learning to fly. Most traditional general-aviation flight schools will tell you that you can learn to fly at your own pace and when you want to. This is true for those schools, but this casual approach is one of the primary causes of the eighty percent drop rate these schools experience. I'm sure you'll agree that this high dropout rate is not good for the students or the schools."

"Also, it is a well proven fact that if in the process of acquiring new knowledge and learning new skills you only infrequently focus on the underlying tasks, it will take you an inordinately long time to reach your learning goals. This has a lot to do with our mind's ability to retain information if it is not frequently reinforced. So, for optimum training efficiency, we have to devote adequate time to learning to fly, and that time has to be scheduled into a reasonable and structured timeline. In fact, if you're not willing to commit to a reasonable timeline for the completion of your training, the Palwaukee Aero Club will not accept your application for membership."

"If you add up all of the directed self-study, ground school, simulator and airplane training time in the Center For Airmanship Excellence Introduction To Personal Flying Program, Sport Pilot Qualification Program and Personal Mobility Pilot Qualification Program, you will find that there are a total of 939 training hours in those programs. We target one year as a reasonable training time to complete all of the programs."

"That means you have to dedicate a little less than twenty hours per week to learning to fly. Our experience is that you can stretch out your flight training to a maximum of eighteen months before you begin to significantly diminish the effectiveness of your training. If you spread the 939 training hours over those eighteen months, you'll have to focus on learning to fly for around twelve hours every week during that period," Josh told them.

"The important thing to remember is that in order to learn to fly right with reasonable effectiveness, you have to commit from around twelve to twenty hours a week over a twelve-to-eighteen month period. Our recommendation is that if you're not willing to make and fund that commitment upfront before you start learning to fly, don't bother. Now, keep in mind, this is the timeline you'll need to commit to if your airmanship goal is to become fully qualified to fly the club's SR22. And I want to emphasize that if you commit yourself to the pursuit of airmanship excellence, that commitment will impel you to also commit to a regime of life-long learning."

"If you decide to only fly the club's Light Sport Aircraft for your personal flying, you only need to dedicate around fifteen hours per week over about a five-month period to get fully qualified to do that. And if your airmanship goal is to be fully qualified to fly the SR22, then by the time you reach this five-month milestone, you'll be able to fly around by yourself or with one passenger in a Light Sport Aircraft while you continue your training."

"There's one last topic that I'd like to address before I wrap up this briefing," Josh said. "As a member of the Palwaukee Aero Club, you will be assigned to a training team that includes you, the Chair of our Training Operations Committee, a Mentor Pilot, a Senior Flight Instructor, at least one Associate Flight Instructor and a Senior Ground Instructor. Your training team will support your training while you are a member of our club by helping you to decide what training you need and then by making sure that you receive the high-quality training that is specified in your Personal Airmanship Development Plan. You won't find this robust approach to flight training at any other general-aviation flight school."

"Well," Josh continued, "that pretty well wraps up my briefing." "If you have any questions, please bring them up in our Q&A session after lunch. I've enjoyed our time together today and I'm looking forward to seeing you around the club and flying with you when you become a member."

"So, thank you for your attention," Josh said. "And I'd like to leave you with one last thought. Learning to fly will change your life and your lifestyle for the better—forever. And you will find that the things you learn in your Airmanship 2.0 training can be applied with considerable effectiveness to other areas of your life. Thank you."

As Diane walked to the front of the classroom, Dan said to Bill, "Their flight training programs are even better than the ones I went through in the Army. If I had it to do over, this is how I'd learn to fly."

Briefing 5: Member Support

When Diane reached the podium, she said, "Our next briefing, and there's only one more after this one, will be on how we support our members in their pursuit of airmanship excellence." "And to give us that briefing, I'd like to introduce Andrew. He's a captain for a regional airline and the Chair of the Center For Airmanship Excellence's Airmanship Outreach Committee. Andrew joined the Palwaukee Aero Club about two years ago. He's also a Mentor Pilot and flight instructor."

Andrew took the podium and said, "I'd like to add my welcome to all of you." "As Diane said, I joined the club a couple of years ago. I did that because I wanted a variety of well-equipped and maintained private aircraft to fly for my business-and-personal-mobility needs. When I take my wife and two kids on trips, I always try to avoid the pain and hassles of airline travel, and a private airplane is a great way to do that. I also want the kind of support I get in my airline flying to be available to me in my personal flying. The Palwaukee Aero Club is the only place I could find that provided the Airmanship 2.0 support I was looking for."

"I'm going to brief you on how the club supports our members in their pursuit of airmanship excellence. In general, the club supports its members by negotiating, signing and overseeing an ADSO Management Agreement with the Center For Airmanship Excellence. The Center then runs our club and all of our flight and training operations for us."

"First of all," Andrew said, "the Center provides us with all of the policies, procedures, systems and tools needed to run the club." "They are incorporated into the Center-produced Palwaukee Aero Club 'Bylaws', 'Aero Club Handbook', 'Flight Operations Handbook', 'Membership Agreement' and

'Aero Club Business Plan'. The Center also recruits, selects, trains and manages all of the managers and support staff needed for club operations."

"This takes a great burden off our members. After all, we joined the club because we want hassle-free, enjoyable personal flying. We don't want to have to manage and staff an organization of this size and complexity as part-time volunteers. And professional management means a more efficient and safer operation. Of course, our members get involved in a lot of volunteer work for the club, but that's because they want to."

"We've found that this approach to managing the club has worked just as the Center For Airmanship Excellence designed it," Andrew said. "The members don't have to get involved in the running of the club if they don't want to, the professional team that makes the club run smoothly on a day-to-day basis does an excellent job and we all feel that we have a say in how the club operates. We also like the business-transparency policies the Center built into the club structure. All in all, I couldn't be more satisfied and pleased with how the club supports me in my pursuit of airmanship excellence."

"The Center's approach to the club's management also yields great dividends in terms of member service. For example, my wife and I always feel like the staff is pampering us whenever we're at the club or flying one of the club's airplanes. Our expectations are always somehow exceeded. The support staff is consistently friendly, helpful, polite and sensitive to the needs and desires of our members."

"They're that way because of how they're recruited, selected and trained. Also, they truly like their jobs and the people they work with, including our members. But I think that the biggest factor that shapes the staff's member-support mentality is the fact that every member of the staff is meticulously selected and trained and they all have a sincere passion for aviation. So, we all feel like one big team here, no matter what our personal status in the club may be. Mutual respect is one of the core values of our Airmanship 2.0 culture."

Andrew continued, "The Center also sets up and manages the leases on all of our aircraft, simulators, support equipment, hangars and clubhouse." "This allows our members to have the latest, best airplanes to fly at very reasonable rates and ensures that the support equipment and facilities we need to run the club are available to us without the club having to raise the capital that would be required to purchase them. And here again, very little involvement is required of our members to make all of this happen."

"There's another thing I want to mention regarding the ADSO Management Agreement we have with the Center," Andrew said. "The precision, efficiency and effectiveness of our club's operations are monitored by the Palwaukee Aero Club Board of Governors. This ensures that our club members are getting what we're paying for."

"The Center provides many other support services under the terms of the ADSO Management Agreement. These include the establishment and nurturing of our Airmanship 2.0 culture. We have an Airmanship 2.0 culture at the airline I fly for. As do all other airlines, the military and some corporate flight departments. The Airmanship 2.0 culture that has evolved over the past thirty years in these organizations has proven itself to be tremendously effective in increasing their level of flight-safety and efficiency. And it has proven to be a great way to support an airman's pursuit of airmanship excellence. The Center For Airmanship Excellence has adapted the Airmanship 2.0 cultures found at the airlines and military for use by personal flyers like us."

"Our Airmanship 2.0 culture includes the shared beliefs, assumptions and values that have proven to be important to an aviation culture that holds flight safety and efficiency as its two highest ideals. Let's take a closer look at how our Airmanship 2.0 culture works. If you ask most general-aviation flight organizations that still practice Airmanship 1.0 if they have an aviation-safety culture, they'll undoubtedly tell you 'yes'. Unfortunately, what most of them mean by an aviation-safety culture falls far short of what our Airmanship 2.0 culture is like. Usually, they're referring to a couple of safety posters tacked up on a bulletin board and occasional admonitions by flight management to 'fly safely'. That's about it. When it comes to developing an aviation-safety culture, as in most things in life, the devil is in the details."

"A real aviation-safety culture has norms and procedures for how everyone in the organization interacts with one another. It also has established procedures for methodical and comprehensive decision making. And it has a robust network of internal communications channels as well as a professional safety-management system." "Let me explain," Andrew said.

"The norms and procedures we use in our Airmanship 2.0 culture are codified in our Flight Operations Manual and our training programs. It's easy for our members to find out how we do things around here, and everyone knows what to expect when we interact with other club members and our support staff. The decision-making procedures we use are based on tried-and-true decision-making principles that have been proven to work in professional Airmanship 2.0 cultures. We learn them in our training programs and we religiously practice them so that they become second nature to us. And we have a large toolbox that's full of validated decision-making techniques like online flight-risk-assessment tools and aeronautical-decision-making checklists."

"Our internal-communications channels include a members-only Website with plenty of safety-related content and a safety forum, and a weekly e-Newsletter. We have a non-punitive, FAA-approved Aviation Safety Action Program that helps us to identify potential flight-safety concerns. This program is fueled by voluntary reports made by our members and support staff. These reports are fed directly into our safety-management system. We also hold a monthly club safety briefing here at the clubhouse. We offer the briefing four times a month so all of our members have a chance to attend. Our members are required to attend at least one of these briefings each month. And then throughout the year, we hold several aviation-safety-related events with a strong social aspect to them. Our members really enjoy these events. They include award banquets, field trips, fly-outs to various aviation facilities and events, seminars and Webinars."

"The safety-management system that is integrated into our Airmanship 2.0 culture is very similar to the one used at my airline. By the way, I'm going to start using another aviation acronym when I refer to a safety-management system. That acronym is S-M-S. The key to an effective SMS is data collection and analysis. First of all, let's look at what kind of data we collect and how we collect it. All of the aircraft that we fly are equipped with an **Appareo Vision 1000** system. This system is essentially like the flight-data recorders used on airliners. I'm sure you've heard the news media refer to them as 'black boxes'. The flight-data-recording unit on our aircraft is much more compact than the ones used by the airlines, but the Appareo system's recorder captures all of the flight data that is important to us plus it captures cockpit video recordings. The Appareo continuously records the aircraft's altitude, airspeed, heading, attitude, accelerations and position. In addition, cockpit video and communications are recorded. All of our aircraft also have engine-performance monitoring and reporting systems In addition to the Appareo Vision 1000."

"When one of our aircraft completes a flight, this data is uploaded to the Appareo server via a Wi-Fi connection as the aircraft approaches our hangar complex. The data is then run through **Appareo's Aircraft Logging and Event Recording for Training and Safety** system. Here's another acronym for you. A-L-E-R-T-S." "We refer to it as 'Alerts'," Andrew told them.

"ALERTS is a Flight Operations Quality Assurance, or F-O-Q-A, application that crunches the data and then outputs the results to the Center For Airmanship Excellence's Safety and Standards Committee in the form of reports that are in various formats. For example, the committee routinely gets an ALERTS report if any aircraft limitations are exceeded during a particular flight. The parameters associated with the exceedance are also reported. The report also shows any trends that may be developing that are associated with this exceedance."

"The Safety and Standards Committee then analyses these reports and takes the appropriate action. This may involve a simple safety alert that is transmitted to our members via the redundant communications links we maintain—like email, text messaging, LinkedIn, Twitter and Facebook—or the committee may specify a new training standard that is designed to ensure that our members have the knowledge and skills they need to handle a particular risk. There are several other courses of action the committee may initiate in response to an ALERTS report. All of them are aimed at improving the safety and efficiency of our club."

"After an appropriate response is made to an ALERTS report, the system and the Safety and Standards Committee monitor the results. If additional actions are needed to remove or mitigate the revealed risk, the committee keeps working the problem in this 'closed-loop' manner. This is exactly how an SMS works at an airline."

"There are several other safety-enhancing aspects built into our SMS," Andrew said. "Among them are the use of a flight-risk-assessment tool before every flight, a policy of shared decision-making, robust checks-and-balances, teamwork and safety-leadership practices. I firmly believe that a well designed and properly run SMS, along with appropriate continuous training, can all but eliminate airman-induced accidents. That's been proven at the airlines and we've proven it here at the club."

"The Center For Airmanship Excellence also manages our aircraft-utilization-optimization plan. Essentially, it allows us to strike the right balance between aircraft availability and efficiency. The availability of our aircraft for our members is very important to them. They're busy people with active schedules. They frequently need to fly on short notice and they often travel on multi-day trips." "If they can't get an airplane when and where they need one, the value of their membership in our club diminishes," Andrew told them.

"If an airplane only flies, oh let's say, four hours a month, then it is obviously available most of the time—about ninety-nine percent availability to put a number to it. On the other hand, the more hours we fly an airplane, the lower our overhead costs are per hour. And the more hours we can fly an airplane in a given month, the fewer of that type aircraft we have to lease. This is also very important to our members. We all want to keep our costs down where possible." "So," Andrew reflected, "we work hard to strike the right balance between availability and utilization."

Andrew projected a pie chart on the classroom's viewing screen. It had two sections. The green one was about five times larger than the smaller red section. The green section was labeled "82%" and the red "18%". Andrew explained, "This pie chart will help you to see what I mean." "For example, we're currently flying our SR22s about forty hours a month on average. Most of our members fly between the hours of 7 AM and 9 PM. That means an airplane is available on average twelve hours in any particular day. That results in a total availability of about 360 hours in a month."

"So, if you do the math, you'll find that the forty hours that the airplane is flown is only eighteen percent of the time that it's available. The other eighty two percent of the time, the aircraft is available for scheduling. We've been using this type of scheduling strategy in flying clubs for decades, and this level of utilization means members usually don't have a problem scheduling an airplane when they want one. And the revenue from the forty hours a month pays all the overhead and operating costs on the airplane at the prices we charge ourselves."

"Availability is also enhanced by the relatively large number of aircraft the club has in its fleet. The most popular airplane in our fleet, in terms of use, is the SR22. We have twenty-one of them. The SR22 is the club airplane that I fly the most and I rarely have trouble scheduling one when I want it. Of course, schedule conflicts do sometimes arise. That's one of the compromises you have to make for much more affordable air travel than if you owned the airplane yourself."

"If a potential schedule conflict does arise, our OpsCenter team does everything it can to provide a range of options to the members who are involved in it. For example, our Planeshare Program allows members to share flight costs by teaming up on flights that serve both members' needs. Here's how it worked for me a couple of months ago. I wanted to fly my wife down to our vacation home in Mountain View, Arkansas for a couple of days of fishing and visiting with friends. As it happened, all of the club's SR22s were booked during that time so the OpsCenter couldn't schedule one for me for the trip. However, when they informed me of that, they also told me that they had arranged for a Planeshare flight with another member who already had an SR22 booked for a business trip to St. Louis on those two days."

"The OpsCenter worked out a deal with him where we shared the airplane and the cost of the flight to and from St. Louis, and I paid for the whole cost for a round-trip flight from St. Louis to Mountain View and back. I dropped him off in St. Louis and picked him up there for the return flight to Chicago Executive. The other member and I had a great time flying together and we both saved money."

"The OpsCenter also had another option for me to consider. The club had arranged for a charter flight in a small jet for another member over the two days

I wanted to be in Mountain View. The other member was flying down to Bentonville, Arkansas for a business meeting. He was leaving the morning I wanted to leave and returning to Chicago Executive late the next day. The OpsCenter arranged to have this member drop my wife and me off and picked up at Mountain View, which is only slightly off the beaten track between Chicago and Bentonville. My cost for the flight was quoted as being slightly less than what it would cost me if I had used the SR22 by myself. However, I chose the Planeshare option because even though I drive an airliner around for my day job, I prefer to fly myself rather than deadhead in the back of an airplane, even if it's a business jet with a luxurious cabin. That's an affliction most airmen share."

Andrew continued, "Another big factor that plays into aircraft availability for our club members is the reciprocal-membership agreements we have with the DuPage Aero Club and the Aurora Aero Club." "These agreements allow members of our club to fly any aircraft that they're qualified to fly in the fleets of those other clubs."

"For example, both of these clubs have a large fleet of Cirrus SR22s that are equipped just like our SR22s. That's because the clubs are both certified by the Center For Airmanship Excellence as Airmanship Development Support Organizations. They also lease their SR22s from the same company as we do. This provides the members of all three clubs with across-the-board fleet commonality."

"So, instead of just having the SR22s in our club's fleet available to them, our members have the combined SR22 fleets of all three clubs available for scheduling. And they can pick up another club's aircraft at the other club's home airport, or the OpsCenter will arranged for it to be sitting on the ramp outside this clubhouse for them. Our members also have access to the DuPage Aero Club and Aurora Aero Club clubhouses."

"This arrangement really works well for the members of all three clubs. In fact, we'd like to have reciprocal-membership agreements with more clubs that are Center For Airmanship Excellence-certified Airmanship Development Support Organizations. I believe that will happen pretty soon as I know of at least one new club that is in the offing. The Center For Airmanship Excellence thinks that there will ultimately be at least seven clubs in the Chicago area. So, as you can see with all of the business-and-personal-mobility solutions Palwaukee Aero Club members have at their disposal, it is very rare when the club can't get us where we want to go in a manner that is not only acceptable to us, but also enjoyable and efficient."

"Many of our members own an airplane that's compatible with the club's fleet." "They either fly enough to justify owning an airplane for their exclusive

211

use, that's around four hundred hours a year, or they just want to have exclusive use of a private airplane," Andrew said. "I was personally bitten by the aircraft-ownership bug a couple of times before I joined the club," Andrew said. "If you're a bug-bitten member of the club, the Center For Airmanship Excellence will help you to select and purchase the type of aircraft that works for you and it will arrange for it to be professionally managed for you. Believe me, if you own a private airplane, you want someone to manage it for you. You really don't want to take on that challenge if you don't have to."

"And if you want to reduce your overall cost of ownership, you can even arrange for the club to lease your airplane when you're not using it. It will be treated just like a club airplane. You've already heard about how we take excellent care of all of our airplanes. So, this option works very well for some of our members who are willing to share their baby on a limited basis with other club members. And some of our members have formed partnerships, with the help of the Center, that purchase airplanes and then lease them to the club."

"If you can't justify owning an airplane for your exclusive use but you still want to enjoy the advantages of personal flying, you'll have to share airplanes with other airmen. There are plusses and minuses to sharing airplanes. The biggest advantage to sharing airplanes is that your overall cost of flying will be much lower than if you have to carry the cost of aircraft ownership all by yourself. Another advantage is that in a shared-aircraft situation, you can usually hand off the trials and tribulations of aircraft management to someone else."

"However, there are some disadvantages to the shared-aircraft arrangement. Chief among them, to my mind, is the fact you don't really know how the other people who fly the airplane treat it. Have they overstressed the airplane in a way that may lead to a mechanical or structural failure? Have they mistreated the engine thereby increasing the chances of unexpected engine failure? Are they flying the airplane in a manner that increases maintenance costs that have to be shared by everyone who is in the shared-aircraft relationship? And will the airplane be available when I want it? When I arrive at the hangar to preflight the airplane, will I be faced with an airplane that is dirty and/or messy? These are problems that are typically faced by aircraft renters and traditional flying club members. In fact, these are the primary reasons I decided to buy my own airplanes before I found out about the Palwaukee Aero Club."

"The Center For Airmanship Excellence has addressed these questions and found good, solid ways to circumvent these problems," Andrew said. "For example, we train all of our pilots on how to fly our airplanes in a manner that doesn't result in questionable mechanical reliability. And, the Appareo Vision

1000 and ALERTS systems, as well as digital-engine-monitoring technology, continuously monitor how all of our aircraft are flown."

"If the pilot of one of our airplanes exceeds its operating and/or best-practices parameters, our monitoring systems let us know about it and we can take appropriate follow-up actions to make sure the airplane is in tip-top shape. We also retrain the pilot who flew the airplane so that it doesn't happen again. We also employ more-frequent maintenance inspections as an additional guard against unknown mishandling of our airplanes. These procedures have completely overcome my fears about having to fly an airplane that has been abused by the pilots who flew it before me. This monitoring and training also result in lower maintenance costs for the club, and that means a lower cost of flying for all of us."

"We've already talked about how we handle aircraft availability issues. So, that shared-aircraft problem is also solved. And you've been briefed on how our ground crew takes care of our airplanes to make sure they're clean inside and out and ready to go for us when we come out to fly. The club handles this issue just like a corporate flight department with very expensive business jets does."

"So, the bottom line is that the Palwaukee Aero Club has eliminated, or at least significantly lessened the impact of, the issues faced by airmen who share airplanes. I honestly don't even think about these issues anymore. However, I would have to if I once again entered into a shared-aircraft relationship that doesn't have the policies and procedures the club has. Unfortunately, I don't know of any other aircraft-rental operation, flying club or aircraft partnership that has these safeguards in place."

"I'd like to expand a little on what you've already heard about how we maintain our airplanes. Our aircraft are maintained by the company that leases them to us." "Let me highlight some of the more-important aspects of how they do that." Andrew said.

"First of all, the leasing company uses a maintenance-management system that has been approved by the Center For Airmanship Excellence. This system makes it relatively easy for their maintenance supervisors, who are all very experienced aviation-maintenance technicians with aircraft-inspection authority from the FAA, to keep track of all the maintenance that is required to keep our airplanes in compliance with FAA regulations and in an excellent airworthiness condition. This system ensures that all required maintenance is done in a timely and efficient manner and nothing drops through the cracks."

"Also, we use what is called a 'progressive-maintenance plan'. Most personal-flying operations don't. They inspect their airplanes only once a year, or in some cases where it is required by the FAA, every one hundred hours of flight time. These so called 'hundred-hour inspections' are done every five or six months in a typical aircraft-rental operation. Under the FAA-approved progressive-maintenance plan the leasing company uses, mechanics inspect our airplanes every two or three days."

"It is obviously better to have an aviation-maintenance technician looking over the airplanes you fly on this frequent schedule rather than only once or a few times a year. However, there is a caution in doing progressive maintenance. The historical data shows that when maintenance technicians open up an airplane for inspection or do maintenance on it, there is a risk that they may leave something loose or disconnected. That problem is solved through a rigorous quality-assurance regime that makes sure that doesn't happen."

"And speaking of maintenance," Andrew added, "we include in every member's Personal Airmanship Development Plan at least one half-day training session per year wherein the member assists an aviation-maintenance technicians with doing work on one of our airplanes. This builds trust between our airmen and our aviation-maintenance technicians and it significantly increases the airman's knowledge of the aircraft he flies."

"The aviation-maintenance technicians who work on our airplanes are all appropriately certified by the FAA and they have all completed maintenance training that is conducted by the manufacturers of the airplanes we fly. They also undergo frequent recurrent training to keep them up-to-date. This is unusual in the personal-flying world."

"Most personal flyers have a rather casual approach to how the airplanes they fly are maintained. I find this hard to understand since a mechanical failure in flight can have such dire consequences. As a professional airman, I take an intense interest in how the airplanes I fly are maintained. I'm completely satisfied with the way our club aircraft are looked after by our maintenance team."

"I also want to take a minute to address how we insure our flight operations," Andrew said. "Specifically, our aircraft-accident insurance. That insurance covers the two major liabilities that can arise from hurting someone in an airplane or on the ground, inflicting damage to property on the ground and/or damage to the aircraft itself. We carry liability insurance for the risks associated with causing damage to persons or property with one of our airplanes. Most personal flyers and personal-flying organizations also carry liability insurance. However, the liability limit in their insurance policies is normally only $1,000,000, and that limit can be further reduced to $100,000 per passenger in the airplane."

"This is one of the biggest concerns I, and many other members of the club, had about flying personal airplanes. The maximum amount of liability damages that an insurance company is willing to pay under the terms of a typical personal flyer's liability policy could very easily be exceeded by the multi-million dollar damages awards that are handed down routinely by judges and juries. And none of us wants to face the prospect of paying the balance of those damages out of our personal funds."

"The Center For Airmanship Excellence discovered a viable solution to this problem by working with one of the best aviation-insurance agents on the planet—Brint Smith. Brint is VP Aviation at John F. Throne & Co. where he serves the aviation risk and insurance needs of very affluent, corporate and charter-management aircraft owners and operators. Previously, Brint designed, built and led the high net worth aviation practice for a major international insurance broker. In this capacity, he provided personal aviation asset protection and liability insulation risk and insurance solutions for individuals and families. Clients have included members of the Forbes 400, managed and fractional aircraft owners, yacht-based aircraft, vintage and flying collectibles, owner-operators of turbine aircraft, professional sports teams, airlines, corporate flight departments, flight-schools, aircraft management companies, helicopter operators, fixed-base operators, airports, and manufacturers across North and South America. He is a single-engine rated, private pilot. He chairs the National Business Aviation Association (NBAA) Tax Committee's Risk Management Working Group and he is a member of NBAA's Aircraft Transactions Working Group. Brint is a Board Member of the Pacific Northwest Business Aviation Association and he speaks frequently on aviation risk and insurance topics and has authored several technical articles."

"Brint helped the Center to put together a consortium of aviation-insurance underwriters who were open minded and concerned about solving two of the most pressing problems faced by the personal-flying community today—the personal-flying accident rate and the rapidly dwindling number of personal flyers. After learning how an Airmanship 2.0 organization like the Palwaukee Aero Club could reduce it's accident rate to near zero, just like the airlines, the consortium decided to join with the Center in tackling this enormous barrier to entry for anyone who wants to become a personal flyer."

"The underwriters agreed to increase the liability limits in any policy owned by a Center For Airmanship Excellence-certified flight organization to a level that is sufficient to remove the liability-limits concern from the minds of its members. And as an added show of confidence in the Airmanship 2.0 approach to personal flying, the underwriters significantly reduced the premiums they charge for liability and aircraft-hull, that means 'collision damage', insurance for Airmanship 2.0 flight organizations. Needless to say, this is a tremendous benefit to our members' pocketbooks and peace of mind. In fact, our members aren't faced with deductibles or any other liabilities if they have an accident while flying one of the club's aircraft. This is not the case in most personal-flying organizations."

"Now let me take this briefing in a little different direction as we near the end of it," Andrew said. "One of the really exciting ways that we support our members' airmanship development is by organizing and offering unique Airmanship Experiences to our members. For example, just last weekend we held a 'Warbird Smorgasbord' here at the club. We brought in four different World War II military aircraft—a **T-6,** a **P-51,** a **B-25** and a **B-17**. During the two days, over fifty of our members attended short ground schools on these airplanes and then went out and flew each one of them. I was fortunate enough to get a highly prized slot to attend the event. And let me tell you, even though I fly jets for a living, I have never been as thrilled and satisfied by anything in my flying career as I was by my experiences flying these warbirds of my dreams."

"Another good example of how the club supports our airmanship development is the group flight last year down to the little town of **Tolhuin** which is located on the southern tip of South America in the Argentinian province of Terra del Fuego. Two of our Mentor Pilots organized and ran the trip for five of our members and their guests. They flew six of the club's SR22s on the trip. Unfortunately, I was unable to join them, but by all accounts the trip was a resounding success. The members who took part in the Airmanship Experience are still exercising their bragging rights around the clubhouse."

"These are but two examples of the Airmanship Experiences that are offered to our members. In fact, I'm the lead Mentor Pilot on a round-the-world flight that our club plans to complete early next year. So far, ten members are signed up for this Airmanship Experience that I think will top them all."

"The club also organizes a variety of volunteer and charity activities for our members. One Saturday each month, we support the Experimental Aircraft Association's **Young Eagles Program** by providing free flights in our club aircraft to young people who want to learn more about aviation. Many of our members belong to the **Volunteer Pilots Association.** This organization provides flights to people in need that must travel to receive necessary medical treatment. Our members fly these folks to and from where they need to go for their medical care in our club aircraft at no charge."

"Also, several of our members fly for **LightHawk**. It's a volunteer organization dedicated to helping conservation groups to survey the earth from the air. They use our club aircraft, usually one of the AirCams, for the flights at no charge to LightHawk. Our members who participate in this program all tell me that the flying is really fun and they feel like they're doing something positive for the environment. The club supports several other charitable organizations with our aviation resources and expertise."

"Our members also support the **FAA Wings Pilot Proficiency Program** in several ways. All of our members are enrolled in the program as part of their Personal Airmanship Development Plan. But several of our members go the extra mile by creating and presenting Wings Program pilot-safety seminars and Webinars as FAA Safety Team volunteers."

"Now, you've been hearing us talk about your Personal Airmanship Development Plan that each of our members creates and maintains along with the help of his or her Mentor Pilot. This plan is the essence of the support that the club provides for our members in their pursuit of airmanship excellence. You've heard about this plan before, but I'd like to go into a little more detail about how it works and what it means to the airmen in our club."

"It is a requirement of continued club membership to maintain an up-to-date and active Personal Airmanship Development Plan. We do this because the effective pursuit of airmanship excellence requires it. We know that it is virtually impossible to continuously improve our airmanship without one. And every member of the Palwaukee Aero Club is committed to the pursuit of airmanship excellence. The many benefits of this pursuit accrue to our members individually and to the club as a whole."

"Personal Airmanship Development Plans are living documents and they can be updated at any time based on the needs or desires of the airman following the plan. The plans are always focused on specific airmanship goals like maintaining the airmanship knowledge and skills needed to sustain an airman's qualifications, or on expanding an airman's airmanship knowledge and skills in particular areas of airmanship. They are based on a twelve-month planning period, and the planning process begins with an assessment of an airman's current airmanship level. A Mentor Pilot makes this assessment by meeting and flying with the airman. The Mentor Pilot and airman use airmanship-assessment tools provided by the Center For Airmanship Excellence to objectively measure the airman's current airmanship knowledge and skills."

"After assessing the airman's current level of airmanship, the airman, with the Mentor Pilot's help, sets realistic airmanship goals for the next twelve months. These specific goals encompass both airmanship-knowledge and airmanship-skills goals. Once the specific airmanship goals have been identified, the Mentor Pilot and airman determine the airmanship training that will be required to meet the stated airmanship goals. The training will probably include directed self-study, formal classroom training, simulator training, flight training and directed practice in a simulator and airplane."

"The next step in the Personal Airmanship Develop Plan planning process is to define and locate the resources the airman will need to work his or her plan. These resources typically include directed-self-study materials, a simulator, an aircraft and a flight instructor. And they might include enrollment in a formal training course."

"Next, the Mentor Pilot and airman define specific objectives to be met in the Personal Airmanship Develop Plan. For example, two objectives that may be identified for the six-month point in the plan could be the passing of an instrument-proficiency check and demonstrated precision in short-field landings that exceeds that required by the qualification standards for the type of flying the airman is doing."

"Then, the airman and Mentor Pilot plan the instruction and training that is required to reach the specific objectives that were defined in the previous step. At this point, they will actually schedule the instruction and training at the appropriate points in the plan. This is a very important step. We've found that by actually laying out a schedule, the airman is much more likely to stay on his or her plan."

"For example," Andrew continued, "let's say the objective of passing an instrument-proficiency check at the end of June will require the scheduling of a simulator, an airplane and a Senior Flight Instructor. So the airman actually has the OpsCenter schedule the resources for him on the last Saturday of June. This is done for every Personal Airmanship Develop Plan objective throughout the year."

"The airman then works his Personal Airmanship Develop Plan. While he's doing that, he's challenging himself, finding his limits and improving his airmanship knowledge and skills. He also keeps an Airmanship Journal that contains his airmanship-related questions and thoughts. His Mentor Pilot is available to him throughout the year to help him stay on track and to recommend any changes in the plan that he thinks are necessary."

"At the end of the twelve-month period, the Mentor Pilot once again evaluates the airman's airmanship level. The airman and Mentor Pilot then cast this assessment against the stated airmanship goals in the plan and determine whether they were met, exceeded or not achieved. The Mentor Pilot and airman then put together another twelve-month Personal Airmanship Develop

Plan that will redress any shortfalls and establish new airmanship-development goals for the next year."

"There's another area of member support that I'd like to touch on briefly," Andrew said. "**The Aircraft Owners and Pilots Association**, or as we like to call it A-O-P-A, is the world's largest pilot organization. It is the main advocate for personal flying in the U.S. AOPA was founded in the United States way back in 1939 at the dawn of personal flying. It's membership numbers somewhere north of 400,000. The organization does a lot of really good work in many areas of personal flying. And they produce some fine aviation periodicals and self-study programs for personal flyers."

"Every member of the Palwaukee Aero Cub is also a member of AOPA. The AOPA dues are included in the Annual Membership Package that we all purchase from the club. You're going to hear more about that in a minute. We support AOPA because of all the good work they do to ensure our members' rights to fly personal aircraft and the support they provide to our members in the form of personal-flying education and information."

"In 2010, the Aircraft Owners and Pilots Association issued a report titled '**The Flight Training Experience**'. The qualitative research behind the report identified sixty-seven discrete attributes that described the optimal flight training experience. Many of these characteristics carried over into what many believe to be the ideal personal-flying experience."

"The report came out about the time that the Center For Airmanship Excellence was completing its own research into why the personal-flying community in the U.S. was dwindling rapidly and what could be done about it. The AOPA report verified the Center's conclusions and gave the folks at the Center who were working on the problem a boost in confidence that they were on the right track to solving it."

"The Center For Airmanship Excellence incorporated the sixty seven attributes of an optimum personal-flying experience, and a few more, into its 'Airmanship Development Support Organization Handbook' that the Center uses in its management of our club. And I can tell you from my perspective that they did a great job." "Our annual member surveys have been showing us that the members of our club feel that they are getting high value from their membership in the Palwaukee Aero Club," Andrew informed them. "Since late 2012, AOPA has also been serving the personal-flying community by helping groups like ours to form flying clubs through its **Flying Club Network**. In fact, Steve and the other founders of the club consulted with the **AOPA Center To Advance The Pilot Community** before they settled on using the Center For Airmanship Excellence's approach to setting up and running a flying club."

"Peer pressure, both positive and negative, is another way in which we support our members' pursuit of airmanship excellence," Andrew said. "The peer pressure our members experience is not just a random response to something. It is part of our Airmanship 2.0 culture. It's one of our beliefs that we need to support each other in our pursuit of airmanship excellence by openly expressing our approval or disapproval of an airmanship behavior exhibited by a member when it is warranted."

"The times when peer pressure is warranted is spelled out in the Center For Airmanship Excellence Airmanship Development Support Organization Handbook. For example, positive peer pressure in the form of sincere encouragement is expected from all members toward a member who has hit a learning plateau or is going through a rough patch in an airmanship-training program. On the other hand, a member is expected to exert appropriate negative peer pressure if he or she observes another member using non-standard operating procedures."

"Another topic that I'd like to touch on," Andrew said, "is how the club supports its members in their pursuit of airmanship excellence through the club's 'Passenger Confidence Training Program'." "Every passenger on a club aircraft, if he or she hasn't been through at least the Introduction To Personal Flying Program, must complete at least Phase I of our Passenger Confidence Training Program. Phase I consists of a short video on passenger safety and protocol, and a fifteen-minute session in a simulator that trains passengers to use simple emergency procedures that will save them if the Pilot In Command is incapacitated. Passengers also receive guidance on how to learn more about the basics of flying. They also receive personal checklists that will help them to prepare for a flight in one of our airplanes and to perform the emergency procedures they learn in this phase."

"Frequent passengers on club aircraft, like spouses, children and business associates, can opt to take Phase II of the training. In this phase, they learn how to assist the Pilot In Command by, for example, keeping the cockpit organized for him, watching for traffic and listening for radio communications. And they can learn how to fly the airplane well enough to get it back on the ground in one piece if they should have to. They will also learn how to make a takeoff so that they can take over in case the pilot becomes incapacitated during this critical phase of flight. Our members tell us that this training program, coupled with all of the safety devices installed on our aircraft, have all but eliminated the fear of small airplanes that many of their passengers had. This goes a long way towards reducing the resistance spouses, children, business associates and other passengers may have to our members flying private airplanes."

"The Stick Buddy Program is another great way the club supports our members' pursuit of airmanship excellence," Andrew told them. "The name of the program comes from the term airline and military pilots use for the pilots they are paired with for two-pilot-operational training. For example, at the airlines the captain and first officer on a training crew are called 'stick buddies'. So our Stick Buddy Program is designed to pair up and train our members who want to, at least occasionally, fly with one of our other members or one of our Flight Pros in a two-pilot-cockpit environment."

"Your continuous-airmanship development depends on your flying experiences as well as education and training. Most professional pilots believe that you get more out of your flying experiences by flying with another well-trained and disciplined pilot than you do when you only fly in a single-pilot environment. And well-trained and disciplined two-pilot crews are generally much safer than a single-pilot operation. However, it can actually be more dangerous for two pilots to fly together if they haven't been trained in two-pilot-crew operations and if they don't use standardized two-pilot-operational procedures when they're flying together."

"To qualify to fly in the Stick Buddy Program, a member has to complete the 'Stick Buddy Initial Training Program' and then take periodic recurrent training in two-pilot operations. The training program consists of ten hours of directed self-study, a three-hour ground school and three hours in a simulator with a Stick Buddy where both pilots practice two-pilot operations."

"The club helps Stick-Buddy-qualified members to find compatible Stick Buddies to fly with. You can have as many Stick Buddies as you want. You communicate with the Stick Buddies in your group through the club's social media tools to arrange opportunities to fly with them. It's up to the Stick Buddies to decide if they want to share the cost of a flight they fly together or if one of them is picking up the tab. As you can see, the Stick Buddy Program opens up airmanship-experience-acquisition opportunities and reduces the overall cost of flying for our members."

"Our members join the club's Stick Buddy Program because they want to experience what it's like to operate like a professional two-pilot crew," Andrew said. "They also like the option of flying with another airman if the flight they're planning is unusually challenging. It also gives our members the chance to fly with airmen of varying experience levels, and it can be more fun to fly with a fellow personal-flying enthusiast. And our members especially like the fact that the Stick Buddy Program significantly enhances the value of their membership in the Palwaukee Aero Club."

"So," Andrew said, "that wraps up my briefing on how the Palwaukee Aero Club supports our members' pursuit of airmanship excellence." Believe it

221

or not, there's more to tell you about this topic, but I'm afraid we're out of time. I hope my briefing gave you a good understanding of how the club supports us and of the club's dedication to helping us to become the best airmen we can be while enjoying ourselves immensely. Thank you for your attention." As Andrew was leaving the podium, Dan, who was sitting next to Bill, leaned over and said, "I don't know about you Bill, but I'm already sold on joining this outfit." Bill responded, "me too."

Diane came back to the front of the classroom and said, "Thank you Andrew." "That was a great briefing. Now, we have one more briefing before we break for lunch. Your briefer is Caroline. She's the Palwaukee Aero Club Manager of Member Services and a Sport Pilot. Caroline is going to brief you on some of the specifics of membership in our club."

Briefing 6: Club Membership

Caroline took the podium and began her briefing by telling the group, "I'd like to add my welcome to all of you for joining us here today and say that I'm sincerely hoping your visit with us has so far been enjoyable and enlightening." "Before I go into our club membership structure and process, I want to briefly add to what Diane told you about me being a Sport Pilot. I think you'll find it illuminating. I've been a member of the Palwaukee Aero Club for a little more than a year now. Actually, my husband joined a little over three years ago, but I was not yet what we refer to as an 'Associate Member'. I'll tell what that is in just a minute."

"My husband joined the club so that he could learn to fly himself on business trips and to take me and the kids on vacations. When he became fully qualified to fly the club's SR22s, one of the first things he did was to invite me to fly up to Madison with him for dinner. Of course, before I could book that date with him, I had to complete Phase I of the club's Passenger Confidence Training Program. I had never flown in a small airplane before and I'm not the world's best airline passenger either."

"After I completed the short course, I felt much better about flying in small airplanes and flying in general. A lot of my fear of flying in a small airplane went away as soon as I found out I could return to earth safely if my pilot conked out. At any rate, when I completed the course, I found myself to be enthusiastically looking forward to my first flight with my husband and my first flight in a private airplane."

"We departed this airport in the late afternoon and arrived in Madison just before sundown. On the trip up, I was mesmerized by the view of the countryside from three thousand feet. I'd looked out the window of an airliner before, but never for very long and it seemed like we were so high that all I could see on the ground were large patterns. This was great! I could make out houses, cars, roads, lakes and even cattle in the fields. I knew right away that the kids were going to just love the view. And I discovered something else on the flight. I gained a whole new measure of respect and pride for my husband. He's a wonderful pilot!"

"We took off from Madison a couple of hours after the sun went down. If I was amazed on the way up to Madison in the daytime, I was well beyond astounded on the way back in the dark. The technology in the SR22's cockpit, especially its enhanced-vision system, reassured me that we would not get lost in the night sky. And once we were airborne and on our way home, well satisfied from the wonderful dinner we had in Madison, I felt like Alice in Wonderland as I looked down on the random patterns of lights on the ground and up at the haphazard scattering of stars in the sky. The Cirrus' engine hummed along smoothly and I felt comfortable in my confidence that if anything went wrong my husband could handle it. And if he for some reason couldn't, I could because of the training the club had given me and the safety equipment that was installed on the airplane."

"We weren't even back on the ground at Chicago Executive before I had decided that this was pretty neat and I wanted to do a lot more of it with my husband and kids. We discussed my reaction to the flight in the car on the way home. My husband suggested that if I was going to be flying with him regularly, and especially if the kids were going to come with us, it would be a good idea for me to complete Phase II of the club's Passenger Confidence Training Program. I quickly agreed with him and suddenly realized that I was looking forward to the challenge of proving I could really land a Cirrus all by myself if I had too."

"Well, as they say, the rest is history," Caroline quipped. "After I completed Phase II of the Passenger Confidence Training Program, I was hooked on learning to fly. The next day I enrolled in the Introduction To Personal Flying Program. When I finished the program and was therefore eligible for membership in the club, I applied for an Associate Membership and I was accepted. I completed the Sport Pilot Qualification Program and I'm now happily flying around in the club's CTLSi and Cub. My kids and my husband love to go flying with me in it. Now, I'm thinking about enrolling in the Personal Mobility Pilot Qualification Program."

"Alright, that's enough about me. Now let me tell you briefly about how our club membership is structured and how you can become a member. First of all, we have four membership classifications—Full Member, Associate Member, Student Member and Candidate Member. All members have access to club aircraft and facilities and the airmanship-training programs that are offered by the club."

"Full Members can vote in club elections and on other issues that are put before the membership for a vote. Full Members also have schedule priority over Associate Members and Student members. This means that if a Full Member and, for example, an Associate member both want to schedule a club aircraft or other club resource for the same time period, the Full Member has schedule priority."

"Also, Full Members can 'bump' Associate Members and Student Members from a particular scheduling slot with forty-eight-hours notice. In other words, if an Associate or Student Member has an airplane scheduled more than forty-hours in the future and a Full Member wants to schedule the airplane during the slot the Associate Member has scheduled it, the Full Member will get the airplane. Please note that the Full Member can't bump the Associate member from the schedule within forty-eight hours of a scheduled slot. I'd also like to point out that this doesn't happen very often because we have so many aircraft in our fleet that at least one is generally available at any given time."

"Full Members also get priority scheduling for any special events that the club puts on. I'm sure you remember Andrew telling you about the Warbird Smorgasbord that he attended. You might also recall that he mentioned that he was lucky to get a slot for the event. Andrew got one of the last reservations only because he's a Full Member. At the time that he made his reservation, there was a waiting list for the event made up of Associate and Student Members. An Associate or Student member would have gotten a reservation only if all the slots had not been taken by Full Members."

"Associate Members have to have a family, business or other personal association with a Full Member to qualify for club membership as an Associate Member." "And they need to be sponsored by that Full Member," Caroline said. "Associate Members can also vote on club elections and affairs. Associate members have schedule priority over Student members. The protocol for exercising that authority is the same as for a Full Member. Associate Members can also bump Student Members from a particular scheduling slot with forty-eight-hours notice. Associate Members have scheduling priority over Student Members for any club special events."

Caroline continued, "All Student Members are actively participating in the Center For Airmanship Excellence's Professional Pilot Preparation Program and they're working for the club either full or part time." "Student Members can vote in club elections and on other issues that are put before the membership for a vote. And, Student Members are at the bottom of the totem pole when it comes to scheduling priorities."

"Membership in the Palwaukee Aero Club is by invitation only. That's because we restrict membership in the club to only those airmen and airwomen who are committed to pursuing airmanship excellence through the practice of Airmanship 2.0. That commitment is formalized in our five-year membership agreement. I'm sure you can understand the reason for doing this. The integrity of our Airmanship 2.0 culture is of paramount importance to us. Therefore, it's not good for the club or for our individual members to have pilots among us who do not share a commitment to the pursuit of airmanship excellence and the club's beliefs, values and norms."

"The club will not extend an invitation to join the club as a Full, Associate or Student Member to a Candidate Member until he or she has completed the club's Introduction To Personal Flying Program." "That's because we believe the Candidate Member and the club should have some time to get to know each other before either of them makes a decision about whether or not the club is right for the Candidate Member."

"Candidate Members," Caroline continued, "have limited access to the club's facilities and resources." A Candidate Membership is only good for thirty days. During that time, the Candidate Member will be expected to complete the Introduction To Personal Flying Program and to participate in appropriate club activities. The Candidate Member will also meet with a Mentor Pilot and at least three other club members. Each of these current club members will evaluate the Candidate Member's understanding of the pursuit of airmanship excellence through the practice of Airmanship 2.0, our club's culture and the candidate's commitment to the club's shared goals. The current members report their conclusions to the club's Airmanship Outreach Committee."

"The Airman Selection Subcommittee of the Airmanship Outreach Committee then evaluates the information they have on the Candidate Member and makes a recommendation to the club's president on whether or not to extend an invitation to join the club to the Candidate Member. If a membership invitation is extended to the Candidate Member, he or she will have thirty days within which to sign a membership agreement. Otherwise, the invitation is withdrawn."

"Full and Associate Members must be at least twenty one years old and Student Members must be between the ages of fourteen and twenty one at the time of their membership application. A Student Membership can be cancelled at any time by the club if the Student Member is not actively participating in the Professional Pilot Preparation Program. The Student Membership expires upon completion of the program."

"In wrapping up this briefing," Caroline said "I'd like to touch on one more topic." "I'm sure you're curious to know how much membership in the Palwaukee Aero Club is going to cost you. After all, no bucks, no Buck Rogers. Right?"

"I'm sure you're thinking that something this good has got to cost a pretty penny. If you're thinking that, you're right. All aviating is costly. That's why those of us who fly look more at the value we receive rather than the cost. Of course, there is the affordability issue, but I've found that for most people who learn to fly, it's the priority they place on their spending decisions rather than on whether or not they can actually pay for it."

"In fact, all of our Student Members pay for their flight training by working for the club. And we pay them only competitive wages which are not what most of us would consider to be high. In my opinion, the value my husband receives from his Full Membership and that I get from my Associate Membership is more than worth what we pay for them. Especially when you consider the fact that we can't even find anything comparable. And I can tell you this, as a wife and a mother I can't put a price on the peace of mind I enjoy because of how the club is structured and run."

"If you decide to enroll in our Introduction To Personal Flying Program, you'll be given a homework assignment that entails visits to other flight schools, aircraft-rental operations and flying clubs in the area. Your homework will task you with finding out all about how those flight organizations operate, what they have to offer and how much it costs to fly with them. You, your classmates and your classroom instructor will compare findings and then sort those findings into oranges-to-oranges evaluations of the relative costs and value of the choices that are available to you."

"For example," Caroline pointed out, "when you take into account the fact that it will probably take you at least twice as many flight hours to earn a Private Pilot Certificate at a traditional flight school as compared to learning with us, it is apparent that training with us could be far less expensive." "And, to that savings you have to add the real value of having modern, well equipped and well maintained airplanes with all of the leading edge safety devices that you really should have on any airplane you fly."

"Another thing to think about when comparing the costs of learning to fly is the fact that over eighty percent of student pilots at a typical traditional general-aviation flight school never complete their training for the Private Pilot certificate. Many of them have invested as much at twenty thousand dollars in their training up to the point where they drop out."

"We don't have that drop out problem in our airmanship-training programs. If you do the math, it becomes apparent that actually completing your training will be far more valuable to you than merely throwing your money away on uncompleted training. Add to that the emotional cost caused by frustration, disappointment and disillusionment and you will, I believe, come to the conclusion that learning to fly with us is a relative bargain."

"And, as you've seen," Caroline said, "after you complete your initial flight training, you have to budget for the costs of recurrent training, upgrade training and transition training. I think it will be clear from your research that because our training programs are much more efficient and effective than those of typical general-aviation flight schools, you'll save a considerable amount of time and money over your aviating career as a Palwaukee Aero Club member."

"And you have to look at the fact that in most aircraft-rental operations and flying clubs, your training will seldom make and keep you fully qualified for the type of flying you're doing. As we've seen, lack of proficiency is the root cause of most personal-flying accidents. How much does an accident cost you versus what you pay for training? It could cost you your life."

"You also have to look at what it's going to cost you to fly the aircraft for your business-and-personal-mobility needs. Here again, we have to look at the time, financial, physical and emotional costs associated with being involved in an aviation accident. Are you better off flying thirty and forty year old airplanes that are maintained on a shoe-string budget, or the new, modern, properly equipped aircraft in our club's fleet? By the time you finish the Introduction To Personal Flying Program, you'll have a clear answer to that question."

"Most of our Full and Associate members write off most, if not all, of their business-flying expenses. This factor further reduces the true cost of personal flying. Of course, you'll have to have your accountant look at this aspect for you since the deductibility of your travel expenses needs to be evaluated on an individual basis."

"The costs associated with being a member of the Palwaukee Aero Club are broken down into two categories," Caroline said. "The first cost is your Membership Agreement Fee, and the second is the cost of your all-inclusive Annual Membership Package Fee. Your Membership Agreement Fee is paid at the initiation of each of your five-year membership agreements and your Annual Membership Package Fee is paid at the beginning of each of the five years that your membership is in effect."

"Your Membership Agreement Fee is based on the type of aircraft you plan to primarily fly. This fee covers the upfront leasing fees that the club has to pay for aircraft, simulator and facility leases. Your Annual Membership Package Fee is based on the type of aircraft you plan to fly, the number of occupied hours you plan to put on the aircraft during the year and any training that is specified in your Personal Airmanship Development Plan."

"Let me define occupied hours for you," Caroline said. "Your member account is debited only for the hours that you are actually flying, or occupying, a club aircraft. For example, if an aircraft that you plan to fly has to be ferried to the airport you have specified as your departure airport in your travel request, you are not charged for the ferry time on the airplane. You only pay for the time you actually fly the aircraft."

"Our members really like this arrangement because it allows them to control their travel costs and they can effectively budget for those costs on an annual basis. There are no unexpected costs other than for rental cars, hotel rooms, meals and other travel related costs. They also like the fact that this fee structure ensures that the club is on a sound financial basis at all times."

"There is, however, some flexibility in the fees paid for the Annual Membership Package. For example, if you discover that you need more aircraft hours in your Annual Membership Package, you can purchase Annual Membership Package Supplements in twenty-five-hour increments. Also, if you have unused occupied hours at the end of your membership year, you can roll them over into the next year."

Caroline wrapped up her briefing by saying, "I'm not going to quote pricing for you because it varies too much depending upon your actual and anticipated circumstances." "However, during your Introduction To Personal Flying Program, you will meet with either me or one of my staff and we will prepare a detailed estimate of your membership fees that is based precisely on your situation."

"I'm sure you'll find that the fees compare favorably with the travel costs you're experiencing right now. And I can assure you that accessing private-aircraft travel through the Palwaukee Aero Club is much more affordable than many of the other options that are available to you. In any case, I'm certain that you will see the value of your membership in the Palwaukee Aero Club."

"Thank you for your time and interest. I'm looking forward to working with you when you become a Candidate Member of the club. Now, I'll turn things back over to Diane."

Diane walked to the front of the classroom and said, "Well, that's it for the briefings we have planned for you today. Next on the agenda is a working lunch. After that we'll reconvene back here for a Q&A session. Why don't you all take a ten-minute break and then meet us in the club's restaurant. You all know where that is, right?" After everyone agreed that they could navigate themselves to the restaurant, Diane said, "Great, see you in the restaurant in ten."

As the group dispersed, John Jr. said to his parents, "I really want to join the club as a Student Member. What do you think?" John Jr.'s father replied, "Well son, this certainly looks like a good opportunity for you. But let's wait to see what else they have to tell us before we make any decisions." His mother said, "I agree." "But I'll tell you this. If what they're telling us about the safety of other flight schools is accurate, this is the only place I'll allow you to take flying lessons." With that, they started moving towards the restrooms and restaurant.

As the classroom quickly emptied, Bill said to Dan, "Caroline kind of dodged the cost issue didn't she?" Dan replied, "Not really." "She's right on about how everyone's business-and-personal-mobility needs are different and I'm sure their Personal Airmanship Development Plans are too. So, it makes sense to customize the pricing to what a member needs rather than just throw around numbers that may or may not represent what it's going to cost me to be a member. I'm willing to wait until I talk with the member-service folks to find out what it's going to cost me. After all, we don't have to make any decisions about joining until after we complete the Introduction to Personal Flying Program, and by then we'll be very familiar with the numbers."

Dan said, "But I can tell you that based on my many years of experience in personal flying, she's putting out the straight skinny about how you really need to take a close look at how other flying clubs operate and what they're going to charge you before you try to figure out if you're getting a good deal or not by joining the Palwaukee Aero Club." Bill replied, "I agree, and that's just what I'm going to do." "By the way, since you're such an experienced pilot, do you think they'll require you to go through the Introduction To Personal Flying Program?"

"Are you kidding?" Dan asked. "I'm looking forward to taking the Discover Personal Flying Course to find out more about how the club operates, how they do their training and how the Cub, CTLSi, SR22 and simulators fly. And I'm really looking forward to soloing the Cub in the Solo Pilot Course. It will be the first time that I've soloed in a taildragger. That's something I've always want to do." With that, Dan and Bill followed the others to the club restaurant.

Working Lunch At The Club

As Dan and Bill walked down the glass-walled hallway leading to the club restaurant, they noticed a neon sign over the door to the restaurant that said "The Briefing Room, Members Only". They entered through the glass doors, stopped and looked around. The main dining room of the restaurant that they had just entered had a vaulted ceiling. Hanging from the ceiling were one-quarter-scale models of all of the types of aircraft in he club's fleet. Large monitors covered the ceiling. They collectively displayed an azure-blue sky with puffy white cumulus clouds drifty lazily across it. Occasionally, a warbird or vintage aircraft would glide across the scene in serene flight.

Along the wall to their left, they could see several booths that ran along the common wall with the OpsCenter. Most of the booths were occupied, but the one closest to them was not. So, Dan and Bill walked over for a closer look. The booth sat four, but instead of the typical bench seats one normally finds in a booth, this one had four seats that had obviously come from an airliner cockpit—two on each side of the table. A thirty-inch monitor was mounted on the back wall of the booth. A video was playing on the monitor. It was describing the latest Airmanship Experience that several members of the club had participated in.

As they took in the booth, Arnie walked over and said, "How do you like our booths?" Dan replied, "I think they're pretty cool." Arnie motioned for Dan and Bill to sit down. Arnie took a seat opposite them.

"With this controller," Arnie said, "I can change what's playing on the monitor." "Let me show you," he said. With that Arnie selected a live feed from one of the club's Cubs that was in the Chicago Executive Airport traffic pattern and said, "I can select this view from the cockpit camera in the Cub, and I can select any of the other four cameras mounted inside and outside the Cub." With that, Arnie cycled through the camera views. "We can also hear what the two pilots are saying on their radio and intercom if we want to," Arnie said.

Arnie continued, "Let me show you a few of the other things we can look at." He tickled the controller and called up a view from a camera mounted on the end of the runway the Cub was now landing on. They could see the Cub coming down final approach. Arnie said, "We have cameras mounted on both ends of every runway here at the airport." "They look down the runways and out on the final approach courses to those runways." Arnie switched to a long view of an airplane rapidly heading down a runway towards the camera. As the airplane lifted into the air, Bill said, "Now that's really neat."

"And here's the view from the camera mounted on the roof of the atrium that looks out over the airport," Arnie said as he switched the feed from the runway camera to the rooftop camera mounted three stories above the clubhouse's lobby. "I can also call up any of the displays from the OpsCenter and several informational videos that describe the club and how we operate. And I can select from a menu of aviation videos."

Chapter 3: Airmanship 2.0 In Practice

"As you can see," Arnie said, "We have booths installed along all four walls of this main dining room." "They're all equipped like this one." As Dan and Bill looked around the main dining room, they saw a glass wall off to their left that looked out over the clubhouse ramp and the airport beyond. The wall on their right was festooned with aviation memorabilia including what appeared to be the front end of a full-sized SR22 mounted to the wall and sticking out into the room.

There appeared to be about twenty tables set in the middle of the main dining room. Club members and their guests who were engaged in lively conversations occupied most of them. The wall opposite from where they were sitting was glass. It provided a view of the club's "Captain's Lounge" that was on the other side of the glass wall. Arnie said, "Well, I think it's time to catch up with the rest of Alpha Flight." "Follow me." With that, they got up and headed for the glass doors that led to the Captain's Lounge.

As they passed through the lounge headed for another glass door on the far side, they could see that the lounge also had a glass wall overlooking the club's ramp and the several airplanes that were parked there. Arnie said, "As you can see, the Captain's Lounge is not very busy right now, but from happy hour on this place can get really packed." There was a large circular bar in the center of the lounge, and booths, just like the ones in the main dining room, were arranged along all four walls. The lounge was aviation themed and it looked like a very comfortable setting for club members to relax in and swap flying stories.

The trio passed through another set of double glass doors and entered another hallway. On their left, they saw doors leading to three rooms. Signs above the doors identified them as the "Wilbur Wright Banquet Room", the "Orville Wright Banquet Room" and the "Glenn Curtiss Banquet Room". Dan noted that the rooms were named for three of aviation's most important pioneers.

As they approached the last door, Arnie motioned Dan and Bill into the room and told them, "The rest of Alpha Flight is waiting for us at tables one and two." "Let's hurry and sit down before we miss lunch." The banquet room had vaulted ceilings with a glass wall at the far end that provided an exciting view of the club's ramp. The room was set with ten large round tables with six chairs each.

As Dan and Bill approached table one, they saw that place cards with their names on them were placed next to each other on the far side of the table. Three of their fellow members of Alpha Flight were already seated at the table. One empty seat remained. Its place card read "Steve". Arnie took a seat at table two with the rest of Alpha Flight.

231

As wait staff served the assembled flights beverages, Steve, the club's president, mounted a small stage set up in the front of the room and strode to a podium emblazoned with the club's logo. Steve tapped a couple of times on the microphone attached to the podium and said, "Ladies and gentlemen, may I please have your attention?" "It's good to see you again. As I'm sure you will recall, I'm Steve and I'm the president of the Palwaukee Aero Club. We met briefly a couple of hours ago in the club's business offices."

"It's good to see that you've made it this far in your quest to find out about our club and how we approach personal flying. I sincerely hope that you've had a good time so far and that you're discovering what you came here to find out." Steve continued, "I'd like to take just a couple of minutes before our banquet-room ground crew takes your lunch orders to tell you a little about this room and the agenda for our working lunch today."

"First of all, we're in the Glen Curtiss room. I'm sure you noticed that we have two other rooms just like it. We use these rooms for banquets, large meetings and receptions. The club itself uses them for club functions several times a week, and any of our members can reserve one for their personal needs. We also make the rooms available to community groups at no cost other than for food and beverages. They love coming out to the airport for some of their regular and special meetings and we consider having them here to be part of our Airmanship Outreach Program. You'd be surprised at how many of our new members come to us from these community organizations."

"Now, here's our flightplan for the rest of your orientation visit with us today. As you're enjoying your lunch, and I promise we'll get to that in just a minute, I'd like to ask you to do two things. First, have an open discussion with your tablemates. I'm sure you've already noticed that there is a Palwaukee Aero Club staff member sitting at every table. Please get to know them. The staff members will facilitate the conversations, but please don't pepper them with questions other than what they do for the club and their aviation background. Rather, I'd like to ask that you hold any questions you may still have about the club until dessert. At that time, the staff member at your table will collect your questions for inclusion in the Q&A session we're going to have after lunch."

"When we've completed our working lunch and everyone has had at least one dessert, we'll take the long way back to your classroom so that you can walk off a few of the delicious calories you're about to consume." Steve raised his arm and pointed to his right and said, "On the other side of this wall, you'll find the club's fitness center." "We're going to take you on a short tour of it to complete your look at all of our club's facilities. Then we'll reverse course and head back to the classroom where you'll have the opportunity to get your questions answered. After that, Diane or one of her membership-services staff will be happy to discuss next steps with you if you're interested in pursuing a membership in our club."

"Well, thank you once again for your interest in the Palwaukee Aero Club. I hope that you will enjoy your lunch and your luncheon conversation. I'll be back to say a few words to you after we finish lunch." With that, Steve left the stage and took his seat at table one after pausing to shake hands with and introduce himself to everyone at the table.

The luncheon conversation was indeed enthusiastic and lively. Everyone learned about their tablemates' interest in the club and their thinking about how they could use the club to satisfy their business-and-personal-mobility needs. As the remains of the main course were being cleared from the tables and dessert and coffee were being served, Steve started the working part of the working lunch by returning to the podium.

"If I can have your attention please," Steve interrupted, "I'd like to make sure that we get all of your questions and concerns out on the table, so to speak." "I'll make sure they get to Diane. She'll organize and present them to the panel of experts who will answer them for you in the Q&A session." With that, Steve returned to Table 1 and collected the questions and concerns of his tablemates. By the time everyone had finished dessert, the group's collective questions and concerns were put down on a list and the duplicates eliminated.

Steve walked back to the podium and called once again for the attention of the group. He said, "I hope you all enjoyed your lunch and the conversations you had." "I want to thank you for taking the time to join us today and I hope that you and the club will turn out to be a good match. Let me leave you with one last thought. As Diane mentioned this morning, we're approaching our membership goal. When we reach it, memberships in the club will only be available if a member resigns from the club. So, as you can imagine, it's going to get pretty difficult to secure a membership. My advice to you is to take this opportunity to find out everything you need to know to make an informed decision about joining us. And if you think the Palwaukee Aero Club is for you, then enroll in the Discover Personal Flying Course as soon as possible."

"In closing," Steve said, "I'd like to thank you once again." "I hope to see you around the club. Now, if you'll turn to your left as you leave this room, you'll find your guides waiting for you. Enjoy your tour of our fitness center."

The Tour: Fitness Center

As the members of all five flights exited the Glenn Curtiss Room and turned left, they saw Josh and Andrew holding open two glass doors that led to a long glass-enclosed hallway. The corridor appeared to be about thirty feet in width. Its high ceiling was composed of glass arches that allowed them to look up and see the sky on this warm summer afternoon. The sky was mostly pure blue with a few white, puffy clouds drifting from left to right. The first sixty feet of the wall on their left was made of river stone with two doors in it that identified the "Airwomen's Locker Room" and the "Airmen's Locker Room".

Diane was standing about fifty feet down the hallway just beyond where the stone wall changed to a glass wall. As the group started to coalesce around Diane, she raised her voice and said, "Ladies and gentlemen, let's start our tour here." "The two doors you just passed on your left lead to our women's and men's locker rooms. We can't go in there right now, but if you're curious as to what they look like, there are pictures of them on our Website. They contain lockers, bathrooms and showers for our members' use. Now, if you'll please follow me?"

As the group moved down the hallway following Diane, they could see a large swimming pool on their left through the glass wall. There was a generous concrete deck surrounding the pool and the whole area was enclosed on four sides by glass walls, one of them being the wall the pool shared with the hallway they were in. They could see the club's ramp and the airport beyond through a glass wall on the far side of the pool. People were swimming laps in the pool and lying about on chaise lounges enjoying a relaxing day in the sun. Several people were obviously deep in conversation as they sat at an array of umbrella tables that were arranged around the periphery of the deck.

Diane led the group another three hundred feet down the hallway until they came to a large, two-story, glass-enclosed rotunda that was about fifty feet across. Diane stopped in the middle of the rotunda. The hallway they were in continued beyond the rotunda and another one branched off at a ninety-degree angle to the left. When everyone caught up with her, Diane said, "We call this the Lindbergh Tower." She spread her arms and lifted them upward and said, "as you can see, the tower is a little over two-stories high." "It has both inside and outside balconies on the second floor with a 360-degree view of the airport and surrounding area."

Everyone in the group looked upward and noticed that there were tables and chairs on both the inside and outside balconies. Small groups of people and individuals sat at the tables and were either talking or just looking out at the airplanes and airport from this "Eagles Nest". The group also noticed that the rotunda was capped with a glass geodesic dome.

Diane lowered her arms and pointed to her left. "If you'll look to your right, you can see that we have a snack bar located here in the Lindbergh Tower for the convenience of our members, their family members and their guests who are enjoying the amenities here in the club's fitness center. If you look further to your right beyond the snack bar, you'll see a large patio with tables and chairs that overlooks a putting green. We call this the Lindbergh Patio. Many of our members like to relax there with a cool refreshment after a flight."

"The entrances to the men's and women's workout rooms are located down this hallway behind me, Diane said. "Here again, I can't take you in there because we respect our members' privacy. But there are some great shots of it on our Website. Since staying fit to fly is an important element of Airmanship 2.0, there are usually several of our members in there working out."

"Now, please follow me down this hallway," Diane said as she pointed to her right and moved off ahead of the group down another glass-enclosed hallway that ran between the swimming pool on the left and tennis courts on the right. "We're going to stop about halfway down this hallway," Diane said, "because I want to tell you about a couple of things." "So, be prepared for a sudden stop and no tailgating!" A chuckle rippled through the group as they followed Diane down the long corridor. When she reached a point a little more than halfway down the passage, Diane said in a raised voice, "Prepare to Stop." A second later, the said, "Stop." "Now, please turn ninety degrees to your left. Everyone did so. They were now facing the pool area. Looking back at the locker rooms and banquet rooms beyond it. Diane said, "Great job and good news." "You have all now passed your first spatial-orientation test. It looks to me that you all would make great airmen." There were smiles and nods of agreement all around.

Diane said, "this hallway runs between the pool area ahead of us and the tennis courts behind us." "If we continue to its end, we'll find the Airman's Pro Shop. You'll have a chance to visit it in just a few minutes. As you can see," she said as she pointed at the pool area, "we have an Olympic-size pool for swimming laps, an adults-only workout pool, a kiddie pool, a whirlpool and a Jacuzzi. There are saunas located in the locker rooms."

"If you'll look up at the top of the wall on the far side of the pool area, you'll see a gently curved arch that extends across the pool area between the glass walls on your left and right. Well, that's not one arch; it's a nested telescopic pool enclosure. I guess the technical term for that would be 'sliding roof'," Diane said whimsically.

"The arched transparent panels are made of twin-wall polycarbonate thermal glazing. They're covered with a special film that photo reacts to an electric current. We can make the roof completely transparent or we can adjust it so that it's fully opaque if we want to keep the sun off us on particularly hot days. And all of the outside glass walls in our clubhouse have the same capability. Obviously, with our sliding roof we can use the pool year-round."

"Now, if you'll make a 180-degree turn to your right I'll tell you about our tennis area," Diane said. There was a moment's hesitation on the part of some of the members of the group as they debated about which was their left and which the right, but they all managed to make the turn with only a few embarrassed giggles. Diane said, "Don't be ashamed if you had a little trouble at first in responding to my directions. This often happens to neophyte airmen. Your flight training will solve that problem for you."

Diane continued her briefing, "As you can see, we have four tennis courts for our members' use." "There are glass walls running down the left and right sides of the courts. Through the left glass wall, you can see the club's ramp and the airport beyond that. If you look at the top of the far wall, the one that the workout rooms are behind, you'll note that we have a sliding roof over the courts just as we do over the pool. This, of course, makes the courts playable all-year round."

"OK," Diane said. "Now please make a 270-degree turn to your left." She waited for the group's reaction. Sure enough, about ten percent of them initiated an immediate 270-degree turn to their left, about two thirds of them hesitated a bit but then executed the proper turn and the rest had to follow the lead of the second group. In the end, all of them were pointed back towards the Lindbergh Tower.

Diane said, "Very good." "You're all getting the hang of what we call 'turns to headings'. You'll do a lot more of them in your airmanship training, and you'll get pretty good at them. Now, please continue straight ahead and stop when you get into the Lindberg Tower."

The group followed Diane's instructions with precision. When everyone was gathered in the Lindbergh Tower, Diane said to them, "OK, let's synchronize our watches." "According to my big pilot's watch, the time is now thirteen hundred. That's 1 PM for you novice airmen. Let's rendezvous back in the same classroom we were in at exactly thirteen forty-five for a Q&A session. In the meantime, please feel free to pick up a beverage here at the snack bar. The drinks are on us. Just show the attendant your visitor badge."

"Also, if you'd like to stroll down to the Airman's Pro Shop, please feel free to do that. If you do, there's a selection of free souvenirs that I'd like you to choose from as a reminder of your visit here with us today. Just show your visitor badge to the shop attendants and they'll be happy to show you where they are. And please feel free to purchase any of the merchandise that we have on hand in the shop." "I'll see you," Diane said as she looked at her watch, "back in the classroom in exactly forty four minutes."

As Diane started walking back to the classroom, the rest of the group started to coalesce into small groups of two and three. Some of them moved on in the direction of the Airman's Pro Shop and some went over to the snack bar to pick up a beverage before following the others to the shop. John Jr. and his parents were in the group that went directly to the Airman's Pro Shop. They were among the first to approach it.

The glass-enclosed hallway between the pool area and tennis courts down which they walked ended in another glass-enclosed rotunda. This one was only one-and-a-half stories high and it was about sixty feet in diameter. It too was topped with a glass geodesic dome. About three quarters of the rotunda extended out onto the club's ramp. The view out the windows made them feel like they were immersed in what was going on at the airport. A neon sign over the entrance to the Airman's Pro Shop read "If You're Going To Fly, Do It Right".

As John Jr. and his parents came through the door to the shop, a cornucopia of pilot paraphernalia greeted their eyes. There were pilot flight bags, hand-held electronic devices, aviator clothing, sunglasses, books, CDs, DVDs and sundry other pilot gear that John's parents couldn't put a name to. John Jr., however, new what every thing was. His eyes lit up like a kid at Christmas. This caused his father to get a precautionary grip on his credit card. He could see what was coming. John Jr. moved swiftly among the racks, bookcases and counters while he pointed out all of the must-have items on his pilot-training checklist. This was obviously not the first time he'd thought about pilot gear.

After a quick negotiation, John Jr. and his father settled on the purchase of only one item on his checklist. It was a book. Its title was "In Pursuit Of Airmanship Excellence". The back cover said it was the foundation story for the Palwaukee Aero Club and the other clubs like it that were springing up across the country. John's father agreed to that purchase because he intended to read it too and he suspected that his wife would also want to read it before placing her son in the hands of the club for his professional-pilot training.

As they were paying for the purchase, the shop attendant asked them if they would like a free souvenir. They all nodded their head yes. The attendant showed them where the items were located and told them to take as many souvenirs as they wanted for themselves and to give to their friends who might be interested in what the club has to offer.

John Jr. selected a keychain with a Cub on it for himself and two pens with the club's logo and information about the club printed on them for a couple of his friends. John's father opted for a coffee mug with a picture of a Cirrus SR22 on it for himself and five balsa-glider kits for his friends and business associates who he new had an interest in learning to fly. The club's logo and information about the club were printed on the model airplanes. John's mother

237

chose a keychain with a CTLSi on it for her and five large postcards with pictures of the club's airplanes and clubhouse on it to send to her friends and family members who she thought might be interested in learning to fly.

John Jr. and his parents left the Airman's Pro Shop and started walking back to the classroom. As they proceeded down the hallway between the pool area and the tennis courts, John Jr. said, "You know, if you let me join the club as a Student Member, you'll have access to this fitness center as my guests." "You guys could save a bundle on those exorbitant fitness-club dues you're paying now. And I could stay in shape real easy too." John Jr.'s mother and father looked at each other out of the corners of their eyes. It was obvious that they were both wondering exactly when their son had become such a smooth salesman.

Shorty after John Jr. and his parents left the Airman's Pro Shop, Dan and Bill walked in. They had been enjoying a cup of coffee on the Lindbergh Patio and they stopped by the shop before returning to the classroom for the Q&A session. As they walked through the door to the shop, Dan said, "Wow!" "I think this is the coolest pilot shop I've ever been in. And look at the selection. This is great. I'll be spending a lot of time here after I become a member of the club." Dan and Bill picked up some souvenirs from the Airman's Pro Shop for themselves and for friends of theirs. Bill bought a book called "Stick and Rudder" on Dan's recommendation. By the time they had finished with their purchases, it was time for them to head for the classroom.

Q&A Session

Diane opened the Q&A session at precisely thirteen forty five by saying, "Thank you all for being punctual." "While we were touring the club's fitness center, the staff has been compiling your questions. As you can see, our panel of experts who are going to answer your questions for you is comprised of our five flight guides—Arnie, Linda, Josh, Andrew and Caroline. I'll direct your questions to them, but if another related question pops up please feel free to ask it at that time. We've budgeted up to one hour for this session. If we don't get all of your questions answered during the session, we'll be happy to answer them for you either immediately after we complete the session or at another time that's convenient for you."

"So, let's get started," Diane said. "The first question is: What happens if my husband passes out while at the controls of a club aircraft while I'm with him? That's a good question. OK panelists, who would like to answer that?"

Caroline leaned a little closer to her microphone and said, "I'd like to take that one Diane." "As I mentioned, when I first started flying with my husband I

Chapter 3: Airmanship 2.0 In Practice

was a non-pilot. But the club's Passenger Confidence Training Program taught me how to get my incapacitated husband, my kids and me safely back on the ground. The course also taught me how to send an emergency satellite message and how to use the satellite phone that's installed in all of our aircraft. After I completed that course, I always felt confident that I could survive that type of situation."

"Thank you Caroline," Diane said, "that's a good answer." "Are there any follow-up questions? No. OK. The next question is: How do I get one of those neat training carrels for my basement?" A low murmur passed through the group and several big smiles and nods of agreement could be seen. Diane said, "That's another good question." "Who wants that one?" Josh raised his hand.

"As the Chair of the club's Training Operations Committee, I guess that falls into my area," Josh began. "That may sound like a far-out idea, but actually all of our members have one of the club's training carrels either at home or at their office. As you'll recall from your tour of the Training Center, our computer-training carrels are equipped with a computer, monitor, desktop simulator and a high-speed Internet connection. They are really great for a lot of the directed self-study work we need to do to qualify and remain qualified to fly our club's aircraft. Every club member receives a training carrel as part of his or her initial membership package."

After scanning the room for any follow-up questions and finding no hands raised, Diane said, "The next question is: How does the club take care of my business-and-personal-mobility needs while I'm getting qualified to fly the club's aircraft?" "Josh, I think this one also falls into your bailiwick. Do you want to take this one?"

With a nod of his head, Josh indicated that he would. He said, "As you will recall, it usually takes our members a year to eighteen months to get fully qualified to use our club's aircraft safety and efficiently for business trips and long personal trips." "By the time you get about halfway through your training, you will be qualified to fly yourself on such trips as long as you take one of our Flight Pros along with you."

"Our professional pilots will keep you safe and assist you in advancing your training while you're on the trip when that's appropriate. Before you get to that point, and in fact at any time that it makes more sense for you, the club will arrange for and manage charter flights for you that are flown by our Flight Pros. These charter flights utilize the club's SR20s, SR22s and DA-42s. The club can also charter a twin-turbo-prop airplane or a business jet for you if you need an airplane with a higher capacity, more speed and/or longer range."

"Thanks Josh," Diane said. "And our next question is: What if I need to take more baggage along on a flight than will fit in the baggage compartment of one of the club's aircraft? That's another excellent question. Who wants to take that one?"

Andrew raised his hand and leaned into his microphone. He said, "I'll take it." "I've run into this problem a few times. One time I was flying my wife and the kids to my parents home in Dallas for a Christmas visit. My wife planned to take a boatload full of presents along with us in addition to luggage for four for a five-day visit. When I ran the weight and balance calculations for the SR22 I was planning on flying on the trip and looked at the space that would be required to hold all that, I decided that weight-wise I couldn't take all of it along with us and still have enough fuel to get to Dallas with only one fuel stop and adequate reserves. And there was no way I could fit everything into the SR22s baggage compartment."

"So, I turned the problem over to the OpsCenter. They solved it by recommending that we take the relatively light Christmas presents along with us in the SR22 and they sent our luggage down to Dallas on a club charter flight in a business jet that was going down the day before we planned to arrive. We had enough room and weight margin on the return flight to take our luggage with us. So, there wasn't a problem there. The OpsCenter even sent a ground-crew member over to our house to pick up our luggage so I didn't have to make two trips to the airport. I thought that was a rather elegant solution to my Christmas-sleigh problem. The OpsCenter has a bag full of tricks that they can use to solve problems like this one. At least one of those tricks will usually get the job done."

"Thank you Andrew," Diane said. "And I'd like to add that once you learn how to pack for a flight in a club aircraft, you will rarely encounter this problem. Now, on to the next question, which is: I have two elderly parents who don't get around too well. I'd like to fly them to their summer home in northern Wisconsin and I'm worried about them getting into and out of an SR22. As a Palwaukee Aero Club member, how would I address this problem?"

Caroline indicated she wanted to answer the question. She turned her microphone on and said, "I can speak to that." "My parents are getting up there in age, and I like to fly them to where they want to go. The club's ground crew has worked out ways of assisting mobility-restricted passengers to enter and exit all of our club's aircraft. You'd be surprised at how good they are at assisting the elderly and those with disabilities with boarding our airplanes. One of the big advantages we have over airline travel is that our passengers don't have to cover vast distances getting to an airplane. They're delivered right to it. And of course, members of the club also have the option of

Chapter 3: Airmanship 2.0 In Practice

chartering a bigger airplane that is easier to get in and out of when the need arises."

"Thank you Caroline," Diane said. "The next question is: What kind of ground-transportation solutions does the club provide for its members? I'm going to ask Linda to take this one because the corporate flight department that she manages also uses some of the same ground-transportation solutions we use. Linda?"

"I'd be delighted to answer that question Diane," Linda said. "First of all, the ground-transportation solutions that the club employs are managed by the OpsCenter. The solutions are tailored to the needs of our members and they depend on the resources that are available to solve a particular ground-transportation problem. Let's start with ground transportation in the Chicago area. Most of our members prefer to drive their personal vehicles from their home or office to the airport, and then return to their home or office in them when they return from a flight. However, as we all know, life is constantly throwing little challenges at us like cars that let us down at critical times and matching the right vehicle to the mission. Or maybe you decide to fly a club aircraft one-way outbound and return from your destination on the airlines. You left your car in the hangar here at Chicago Executive and you're now at the airline terminal at O'Hare. How do you get back to your car?"

"Well," Linda said, "the OpsCenter makes life easier for us in this and similar situations." "For example, let's say that it's a cold Chicago morning and you discover that your car won't start when you attempt to leave for the airport to make an early morning departure in an SR22 that you're flying. You're planning to fly to Duluth for a very important business meeting. What are you going to do? The answer is easy. You're going to call the OpsCenter."

"The OpsCenter will immediately dispatch the club's crew van to pick you up if it's available, or they'll send a limo for you. And while you're being rescued, the OpsCenter will, if necessary, re-plan your flight for you and make any required changes to your travel plans. And since you're traveling on your schedule and not an airline's, it doesn't make any real difference if you run a little behind plan. All you have to do is relax, grab another cup of coffee and go with the flow. The OpsCenter handles all the hassles. And you can rest assured that a limo will be waiting for you when you return from your flight to take you home."

"Or," Linda continued, "let's say you need to move your trouble-shooting team of five people and you from your office to the airport, and they just won't fit into your Mercedes SL-Class Roadster." "No problem. The OpsCenter will dispatch the club's crew van or a limo to pick you up. And they'll set up a limo to take you all back to your office when you return. And in the case of being at

O'Hare with your car here in one of our hangars that I mentioned earlier, I'm sure you're already ahead of me on the solution to that problem." "The OpsCenter will send the crew van or a limo to pick you up and bring you back here," Linda said. "The OpsCenter can solve these types of ground-transportation problems for you, and even more challenging ones, no matter where you are in the country 24/7. And, the OpsCenter will even take care of ground-transportation needs you may have that are not directly related to flying a club aircraft."

Diane thanked Linda for her comprehensive answer and said, "I want to direct this next question to Arnie." "Arnie, what if a member wants, or has to, drop out of the club before his or her five-year commitment is up? Will he or she get a refund?"

Arnie pulled his microphone closer to him and began, "I'd be happy to answer that question." "First, let's clarify a couple of things. New members pay a Membership Agreement Fee when they sign their five-year membership agreement. And at the same time, they pay the Annual Membership Package Fee for the first year of the five-year agreement. Then they pay an Annual Membership Package Fee at the beginning of each of the remaining four years of their membership agreement."

"As I mentioned in my briefing, the Membership Agreement Fee covers a member's share of the upfront funding on the leases for the aircraft, simulators, support equipment and the clubhouse. Therefore, the Membership Agreement Fee is a 'sunk cost'. In other words, it can't be refunded if a member drops out because it has already been spent."

"A portion of the Annual Membership Package Fee covers the lease payments on the aircraft and simulators the member flies, and the lease payments on the clubhouse and hangars. As you will recall, all of our leases are structured on five-year terms. The company that leases these assets to us needs to be certain that the lease payments will be forthcoming. Therefore, members have a contractual obligation to pay their Annual Membership Package Fee for the full five-year term of their membership agreement."

"However," Arnie continued, "we've been able to build some flexibility into our membership plan." "For example, if a member needs to drop out of the club for personal or business reasons, the club will remarket the member's membership for him. In this case, the resigning member will be refunded a portion of his Membership Agreement Fee. The amount of the refund will depend on how many years are left on the member's five-year membership agreement, the club's ability to remarket the membership, what the club can resell the membership for and the remarketing fee the club will charge. And, a prorated share of the member's Annual Membership Package Fee for the current year will be refunded to the resigning member when his membership is sold."

"Please keep in mind that a member is contractually obligated to continue to pay his Annual Membership Package Fee until his membership has been resold or the agreement runs out. This refund policy is obviously not perfect from an individual member's point of view. It was designed with the best interests of all of the members of the club in mind." "Obviously, if we followed a no-questions-asked, full-refund policy, the other members of the club would have to make up the shortfall in the club's ability to make its lease payments, and the club would have to tap a non-existent capital pool to cover a refund of the Membership Agreement Fee," Arnie concluded.

"Now that I've said all that," Arnie continued, "I want to point out that our club's dropout rate has been on the order of one percent per year." "The club has had no difficulty in remarketing the memberships of those few members who have had to drop out. So, as it turns out, our refund policy has worked out well for everyone involved. And given the fact that we're getting close to full membership, it shouldn't be much of a problem remarketing a membership to someone on the waiting list."

"Thank you Arnie," Diane said. "Our next question is: Where does the club get its airplanes and simulators?" "Linda, why don't you take this one if you would please?" Linda tapped on her microphone to make sure it was on and said, "I'd be happy to." "We lease all of the physical assets that we need to keep the club running from one company—AvWorld FleetPartners. AvWorld FleetPartners was formed specifically to provide these types of assets to the flying organizations that are certified by the Center For Airmanship Excellence as Airmanship Development Support Organizations, or ADSOs. AvWorld FleetPartners sells Equipment Trust Certificates to investors to raise the money it needs to purchase those assets and then it leases the assets—airplanes, simulators, support equipment and clubhouses—to CFAE-certified ADSOs."

"In most cases, AvWorld FleetPartners gets a quantity discount on the purchase of these assets. This reduces our lease costs considerably. And, the club doesn't have to be continuously involved in aircraft acquisition and disposition deals. This reduces the risk for our members that a valuable asset like an SR22 will have an 'upside-down' residual value when the airplane cycles out of our fleet at the end of five years. AvWorld FleetPartners also manages aircraft that have been purchased by our members and leases those airplanes to our club if the member wants to do that."

"This approach to acquiring these assets relieves the club from having to raise capital and financing for them. This makes it easier and less expensive to start up and run a club. About the only downside to this approach is the fact that we have to commit all of our members to a five-year membership agreement to satisfy the investors who purchase the Equipment Trust Certificates. But, I should point out that this is a case of 'what goes around comes around'. Most of the Equipment Trust Certificates are purchased by club members, although they are under no obligation to do so, because they are relatively secure investments with an attractive rate of return."

"Thanks for covering that question so well," Diane said. "Now, let's move on to the next question. How does the club recruit new members? Andrew, since you're the chairman of the Center For Airmanship Excellence's Airmanship Outreach Committee, perhaps you'd like to take this one."

"Sure," Andrew said as he leaned forward in his seat. "Our New Member Recruitment Plan is managed by the Center For Airmanship Excellence under the terms of the ADSO Management Agreement we have with the Center. We describe the plan as being a highly integrated, high-touch/high-tech, leading-edge referral program. Let me briefly break that down for you."

"Membership recruitment for the Palwaukee Aero Club usually starts with a referral. Referrals generally come from a current club member because our club culture is a referral culture. All of our members actively look for people who are thinking about learning to fly. If they believe an aspiring airman will make a good club member, they use the tools and techniques we provide them with to refer that person to our club."

"Sometimes referrals come from an Internet search engine. And, The Center For Airmanship Excellence refers a lot of potential members to us through its Airmanship Ambassador Program. Often, other non-member pilots who know about us will tell their friends who are thinking about learning to fly to look into a membership in our club."

"And there are a lot of people out there who have been thinking about learning to fly for a long time. Market studies have shown that over fifty percent of the people in advanced and advancing countries have thought about it. Some of them want to learn to fly just to see if they can do it. But most of them want to learn to fly so they can use a private aircraft for their business and/or personal transportation. There are strong indications that about twenty percent of these folks, that's about ten percent of the total adult population in the U.S., or around twenty-million people, will learn to fly if they can find out how to do it safely, effectively and affordably. In any event, you can believe that we receive far more referrals than we have available memberships for. That's one of the reasons our club is expanding so rapidly."

"Most of the people who are referred to us go to the club's Website first. Our site is an information source that provides prospective members with abundant information that allows them to feel comfortable enough with who we are and how we do things around here that they're motivated to contact us. When they do, Diane or one of her staff talks with them on the phone to answer any remaining questions the prospective member may have. The club representative will also probe the prospective member for information that will tell us whether or not he or she might be a good fit with us. If it looks like they might be, they're invited to attend a free orientation briefing like this one."

"After they attend the orientation briefing, prospective members who want to move forward with exploring whether or not a membership in our club is for them will meet with an Airmanship Outreach Committee representative to make sure all of their current questions and concerns have been addressed. The next step for the prospective member is to enroll in our Introduction To Personal Flying Program. Upon completion of that program, the prospective member and the club will, if appropriate, enter into a membership agreement," Andrew concluded.

Diane said, "Thank you Andrew." "I want to direct this next question to Josh who is the chairman of the club's Training Operations Committee. Josh, here it is: What kind of support can I expect from the club while I'm learning to fly?"

Josh cleared his throat and said, "No problem Diane, I can answer that." "To start off, I'd like to remind you that in my briefing I talked a little about your training team. Every club member has a training team that supports him or her in various ways as they move through our training programs. As you will recall, the team is made up of a Mentor Pilot, a Senior Flight Instructor, at least one Associate Flight Instructor, a Senior Ground Instructor and the chair of the club's Training Operations Committee—that's me."

"Your Mentor Pilot makes sure that your training plan is appropriate to your airmanship goals and that it's fully integrated into your Personal Airmanship Development Plan. Your Mentor Pilot also monitors your training progress to make sure you're progressing at a suitable pace, and he acts as your advocate with the Center For Airmanship Excellence. As you will recall, our members' airmanship training is conducted by the Center."

"Your Senior Flight Instructor oversees all of your training and provides you with the majority of your flight training. The Associate Flight Instructors on your training team assist your Senior Flight Instructor with your flight training when assigned to do so by your Senior Flight Instructor. The Senior Ground Instructor on your team makes sure that the ground instruction you are receiving is working for you. I, or one of my representatives, act as the management interface on your team."

"In addition to all of this support," Andrew said, "there is always at least one flight instructor on duty in the club's training center to answer any questions you may have or to help you with running a simulator or a computer-training carrel. The instructor is also available to our members online and by telephone while he's on duty in the training center. We staff the training center from six in the morning until midnight seven days a week. So, you can see that no matter where you are, you always have a flight instructor available to answer your questions."

"The club also supports you in your flight training through interaction with other club members. Every training course starts with an in-classroom ground school that's held here in our training center. The ground school is attended by all of the members who are starting out in the course. In addition to the interaction you'll have with them during the class, you will all be encouraged to talk with the other enrollees in the course online and on the phone as you go through the course. This provides you with a rich mixture of information and alternate viewpoints that will enhance your learning experience. And, it won't take you long to meet many other club members who you can call on for advice and counsel."

"OK," Diane said, "now for one final question: Who does the club charter aircraft from?" "For the answer to that question, let's turn to Arnie." Arnie turned his microphone on and said, "I'll be happy to answer that for you Diane." "The Center For Airmanship Excellence, under the terms of the ADSO Management Agreement we have with it, manages the charter contract we have with a company called AvWorld Connect. AvWorld Connect is a Public Charter Operator that is authorized by the U.S. Department of Transportation to provide air transportation to the public. AvWorld Connect contracts with what the DOT calls 'Direct Air Carriers' for the airplanes and crews that fly its flights."

"The DOT and the FAA certificate the Direct Air Carriers and oversee their operations. The Center For Airmanship Excellence also certifies the Direct Air Carriers that AvWorld Connect uses for its flights. To attain Center For Airmanship Excellence certification, a Direct Air Carrier must have an Airmanship 2.0 culture and operate in compliance with the principles of Airmanship 2.0. Also, the Direct Air Carrier's pilots who fly charter flights for us must be certified by the Center as Flight Pros."

"When one of our members submits a travel request to the club's OpsCenter, aircraft-charter solutions are included in the range of options the OpsCenter presents him with in response to his travel request. For example, let's say our member is traveling with two other people and his profile stipulates that he always wants to be presented with charter options that include both small and mid-sized jets. In this case, one of the aircraft-charter options presented to the member is to charter a light jet like an **Embraer Phenom 100** and the other aircraft-charter option is to charter a **Bombardier Learjet 85**. The Phenom 100 is the smaller and slower of the two. On the other hand, the Learjet 85 is bigger and faster and it has a more-comfortable cabin than the Phenom 100. The Learjet costs considerably more to charter than the Phenom." "For the purpose of illustration, let's say our member decides to charter the Learjet 85," Arnie said.

"When the member let's the OpsCenter know that he's decided to charter the Learjet 85, the OpsCenter gets the ball rolling. It schedules that charter flight with AvWorld Connect. AvWorld Connect then charters the airplane from one of the Direct Air Carriers that it has an agreement with. The OpsCenter also sets up all of the groundside arrangements that were stipulated in the members travel request. The OpsCenter then monitors everything that has to do with that flight until it is completed. If changes are made to the travel plan, and there are usually changes, the OpsCenter manages all of that too."

"So, you can see," Arnie said, "that this approach to providing a wide range of aircraft-charter options to our members is very clean and efficient." "And our members avoid the extra expense and hassles that can be associated with setting up their own charter. The club doesn't have to own or lease expensive business jets, and it doesn't have the costs and problems that are associated with an in-house charter department. It works very well for our members."

"I'd like to mention one other business-and-personal-mobility solution that the OpsCenter can arrange for our members," Arnie continued. "I don't think it's been mentioned before. That solution is airline travel. Although our members primarily use private aircraft to solve their business-and-personal-mobility problems, sometimes airline travel is actually a better solution for all or part of a member's travel itinerary. If that's the case, the OpsCenter provides that solution as an option in response to a member's travel request. If the member opts to use that airline travel, the OpsCenter makes all of the arrangements for the member. You can think of this as an in-house-travel-agency solution that's included in your Annual Membership Package fee."

"Thank you Arnie," Diane said. "I don't have any more questions on my list. So, I'd like to conclude this Q&A session by saying that it was a pleasure getting to know each and every one of you today. I hope you enjoyed it as much as I did. And I know I'm speaking for the other Palwaukee Aero Club staff and members who worked with us today."

"Now, let me suggest some next steps. If you're interested in becoming a Student Member in our club, your next step is to attend the student-parent briefing that will be held right here in this classroom after the short break we're going to take in a minute. The briefing will fill you in on more of the details of how our Professional Pilot Preparation Program works. For the rest of you who came here today to find out more about the Palwaukee Aero Club, I'd like ask you to meet me, or one of my staff members, after the break in the Glenn Curtis Banquet Room where we had lunch. We've reset the room with small tables so that you can meet one-on-one with a club member to discuss where you want to go next. Thank you once again for your interest in the Palwaukee Aero Club. I'm hoping to see all of you around the club after you become members."

Back to Reality

We'll, that's my vision of Airmanship 2.0 in practice. Is it a vision of how you want to pursue airmanship excellence too? I sincerely hope so. However, as I write this in Fall-2013, it is still just a vision. A rather well thought out vision to be sure, but still only a vision. There are several very credible airmen who are right now working to make this vision a reality. I truly believe that if those of us who want the kind of flying experience described above work together, we can make it a reality. If you'd like to join us, you'll find out how in the next chapter.

Chapter 4: *Your Pursuit of More Enjoyable Personal Flying*

Flying might not be all plain sailing, but the fun of it is worth the price.
—*Amelia Earhart*

Now that you've demonstrated your interest in pursuing airmanship excellence by getting this far in the book, I'd like to pass along some tips and ideas on how to go about continuing your quest. But before I do, let's recap some of the more salient points that were addressed in the previous chapters of this book. I think you'll find the short review to be helpful.

In *Chapter 1: My Pursuit of More Enjoyable Personal Flying*, I told you the story of my pursuit of airmanship excellence that has taken place over the last fifty-plus years. I offered this chapter as an example of how the pursuit of airmanship excellence can provide the rewards every current and future pilot is looking for. Those rewards include expanded personal-mobility options, an enhanced lifestyle, increased professional productivity, ever-expanding personal capabilities, a new perspective on life, the satisfaction that comes from mastering new challenges and recognition as a member of a very elite community.

In *Chapter 2: Airmanship 2.0*, I said, "My hope is that when you finish this book, you'll be convinced that there are really only two good choices that a responsible aviator can make with respect to flying airplanes." "You can either pursue airmanship excellence through the practice of Airmanship 2.0, or you can quit flying." And I backed up this rather black-and-white statement with the words of Gen. Yeager: "If you're going to fly, do it right!"

As a personal flyer or hopeful personal flyer, only you can make that choice. In my informed opinion, the FAA is not going to mandate that you migrate from Airmanship 1.0 to Airmanship 2.0—at least not within the next several years. However, as I stated earlier, overwhelming evidence from airline and military flight operations indicates that the practice of Airmanship 2.0 by personal flyers could significantly reduce the current personal-flying accident rate—a goal that the NTSB, FAA and the entire general-aviation industry aspires to.

In my opinion, the aviation-insurance underwriters will probably start requiring policyholders to adopt Airmanship 2.0 as a condition of insurance and/or lower rates at some point down the road. But, I don't think this will happen until personal flyers start practicing Airmanship 2.0 on a larger scale and the benefits become apparent. On the other hand, I do believe that a growing number of flight schools, aircraft-rental operations and flying clubs will embrace it as the diverse benefits of practicing Airmanship 2.0 become obvious to increasing numbers of personal flyers. For now, however, I don't see anyone or any organization or institution that is going to force you to practice Airmanship 2.0 in your personal flying. It is entirely up to you to make what could easily turn out to be a life-and-death decision.

So, have you made the choice to pursue airmanship excellence through the practice of Airmanship 2.0? If not, then before you next leave the ground as a Pilot In Command, I would strongly suggest that you read "Redefining Airmanship", "Flight Discipline", "Controlling Pilot Error", "Blue Threat" and "Darker Shades of Blue", all written by Dr./Col. Tony Kern who I talked about earlier in this book. If the examples and arguments I've presented in this book haven't convinced you to pursue airmanship excellence through the practice of Airmanship 2.0, then you owe it to yourself, your loved ones and your passengers to at least find out what other credible aviators have to say about it. Then, if you're still not convinced, you have a very tough decision to make. Do you continue to fly with inferior airmanship knowledge and skills, knowing full well that you are putting people who trust you at more risk than is necessary; or do you walk away from flying airplanes until you can do it right? In other words, are you going to do it right, do it wrong, or stop doing it altogether?

In this chapter, I also said that it was my hope that we could achieve alignment on the definition of airmanship excellence that I offered:

Airmanship excellence is the possession of airmanship knowledge and skills that exceed those demanded by the type of flying being undertaken.

Are we aligned on this definition? If not, what alternative definition would you offer? If you have one, please send it to me **via email**. Airmanship excellence is a moving target. We need to continuously define it as accurately as we can while at the same time encompassing all the relevant aspects of it. If you are aligned with me on the above definition of airmanship excellence, then are you also aligned with me on how to go about pursuing it as I have described? If not, please let me know what you're thinking. I'm always looking for new ideas that can make the pursuit of airmanship excellence more effective, efficient and fun.

In Chapter 2, we also discussed the differences between Airmanship 1.0 and Airmanship 2.0. I said that Airmanship 1.0 originated in the times of the ancient Greeks and that all aviators — airline, military and general-aviation — adhered to its principles from then until the early 1980s. At that point, airline and military pilots established a new branch of airmanship that I call "Airmanship 2.0". I further asserted that I believe that personal flyers failed to make the turn with the professional pilots in the early 1980s, and that personal flyers are still operating under the principles of Airmanship 1.0.

I also talked about Airmanship 2.0 being a new airmanship paradigm for personal flyers, although it is based on tried-and-true airmanship principles. And I also pointed out that the flavor of Airmanship 2.0 that I advocate for your consumption has adapted the Airmanship 2.0 practiced by airline and military pilots for use by personal flyers. I also said that I don't believe that anyone can effectively pursue airmanship excellence without the support of an Airmanship Development Support Organization (ADSO), and a Personal Airmanship Development Plan (PADP). And I pointed out that the pursuit of airmanship excellence that I am recommending will make you a safer, more efficient pilot and significantly add to the enjoyment and value that you get out of personal flying.

I also pointed out that Airmanship 2.0 continuously enhances your airmanship skills, knowledge and capabilities. Airmanship 1.0 does not. Airmanship 2.0 is a sure-fire way for you to become a safer, more-proficient aviator. And Airmanship 2.0 includes your formal flight-and-ground training, your informal-airmanship training and your PADP. It also includes the types of aircraft you fly and the way that you fly them. We also looked at the Kern Airmanship Model with its three bedrock principles (Flight Discipline, Skill and Proficiency), five Pillars of Knowledge (Self, Aircraft, Team, Environment and Risk) and two Capstone Outcomes (Situational Awareness and Judgment). We use the Kern Airmanship Model within the context of Airmanship 2.0 to structure a Personal Airmanship Development Plan that provides the flightplan for your pursuit of airmanship excellence.

We also studied what I call the "Airmanship 2.0 Model". It consists of three Bedrock Principles (Commitment, Integrity and Shared Experience), five Pillars of Support (Airmanship 2.0 Culture, Safety Management System, Airmanship 2.0 Training, Modern Aircraft and Dedicated Airmen) and two Capstone Outcomes (Safety and Enjoyment). This model can help us to understand how Airmanship 2.0 can be implemented in the personal-flying community.

In the section on Airmanship 2.0 Culture, we took a detailed look at cultures in general and an Airmanship 2.0 culture in particular. We discovered that we all operate within multiple cultures: family, church, work, professional and national. All of these cultures influence us in the cockpit. An Airmanship 2.0 Culture can be viewed as both a professional culture and an organizational culture. We also learned that an Airmanship 2.0 culture is a salient feature of an Airmanship Development Support Organization (ADSO) that supports airmen who practice Airmanship 2.0. This culture includes airmanship-development leadership, a safety-management system, a focus on safety, and values, behaviors and artifacts that promote airmanship excellence.

This chapter also defined what an Airmanship 2.0 culture is and why it was designed that way. Airmanship 2.0 cultures have proven to be effective in radically improving airmanship whenever they have been wholeheartedly implemented. I firmly believe that personal flyers that opt to join an Airmanship 2.0 Culture will have a much safer, richer and more enjoyable personal-flying experience.

Chapter 3: Airmanship 2.0 In Practice provided you with a fictional story of how Airmanship 2.0 might work in the real world of personal flying. Although the story is fictional, it is based on fact and a serious analysis of how we can implement the ideas presented in this book. I personally believe that we can turn the fictional story into reality.

Now, let's take a focused look at your pursuit of more enjoyable personal flying. I will, for the sake of the remainder of this discussion, assume that you have already made the decision to pursue airmanship excellence through the practice of Airmanship 2.0. Here are some tips and ideas on how you can do that.

1. First and foremost, I suggest that you find or create a community of likeminded aviators and/or aspiring aviators who are now, or who you think might be, interested in pursuing airmanship excellence through the practice of Airmanship 2.0. As I mentioned earlier, I don't believe that one can effectively pursue airmanship excellence all by oneself. You need other pilots around you who are also pursuing airmanship excellence. They will provide their support, guidance and understanding for your pursuit of airmanship excellence. You also need them to help you accurately assess your airmanship capabilities and shortcomings. Research shows that human beings are very poor at self-assessment.

2. Also, I firmly believe that Airmanship 2.0's "magic sauce" is the peer pressure, both positive and negative, that runs throughout an Airmanship 2.0 culture. The structured peer pressure of Airmanship 2.0 helps everyone in the group to encourage positive behaviors and discourage anti-Airmanship 2.0 behaviors. We all need well-founded and properly directed peer pressure to help us to excel.

3. Make sure that every, and I mean EVERY, member of your group knows how to go about pursuing airmanship excellence through the practice of Airmanship 2.0, and that they are continuously committed to practicing it. If there are, as Kern calls them, "rogue pilots" in your group, you run the very real risk of succumbing to their way of flying or of your group becoming polluted by the ideas of these non-conforming pilots. The underlying principle here is: If you're going to fly, do it right. And, doing it right is practicing Airmanship 2.0.

4. Always fly with the support of an Airmanship Development Support Organization (ADSO) that has been certified by the Center For

Airmanship Excellence (CFAE). You can't enjoy the full benefits of Airmanship 2.0 unless you belong to a flight organization that is structured and run along the lines of the ADSO I described in this book. Unfortunately, as I write these lines, there are no CFAE-certified ADSOs. One option you have is to try to convert the flight organization you are currently flying with into an ADSO. Another option is to sign up to participate in the Airmanship Development Support Organization Demonstrator (ADSOD) project that CFAE is currently working towards standing up in the Chicago area.

5. If you choose the first option, the Center For Airmanship Excellence will work with you to move your current flight organization from Airmanship 1.0 thinking to an Airmanship 2.0 culture. CFAE will also assist your organization to qualify for ADSO certification. You can contact me **via email** and I'll see that you get the assistance you need. If you elect to join the CFAE crew in its ADSOD project, complete and submit an **ADSOD Pilot Participant Application**.

6. And finally, after you have integrated yourself into an Airmanship 2.0 culture, follow the principles of Airmanship 2.0 with integrity as you pursue airmanship excellence.

If you follow the above advice, your pursuit of airmanship excellence will be both rewarding and effective. You will be able to achieve and maintain airmanship excellence for the remainder of your flying career. Airmanship 2.0 is working every day for tens-of-thousands of professional aviators. I assure you that it will also work for you as a personal flyer.

Acknowledgments

The inspiration that got and kept me going on this book came from my concern for the current state of the art of airmanship among most personal flyers. I am a personal flyer and I count as friends, associates and acquaintances many other personal flyers. My intention is to do everything I can to make Airmanship 2.0 available to every personal flyer on the planet.

I would like to thank all of the personal flyers who attended my FAA Wings Program Webinars and seminars. They freely and enthusiastically gave me their feedback on my proposals on implementing Airmanship 2.0 for personal flyers. And I'd also like to thank the many professional pilots who spent their time listening to my ideas and then sharing with me their opinions on how to bring Airmanship 2.0 to the personal-flying community.

I would also like to particularly thank Capt. H. David Greenberg, Dr. Frank Bacon, Mr. David Shadle, Capt. William Brand, Capt. Richard Sternal, Mr. John Keiper and Mr. Samuel Heiter for always being available to me over the past three years to answer my questions and share their ideas about how we can make personal flying safer, more affordable and more enjoyable. Not only did I thoroughly enjoy our discussions, but I also profited greatly from their insights and expertise.

I also want to acknowledge the untold number of airmen, airwomen, aeromedical doctors and scientists, psychologists and aviation-human-factors experts who laid the groundwork for Airmanship 2.0 through their sacrifices, hard work and brilliant thinking. I, and every other airman who practices Airmanship 2.0, owe them all a debt that can be repaid only by passing the right way to fly along to others. Foremost among this group of pioneers is Dr./Col. Tony Kern. I would like to acknowledge that his seminal airmanship book "Redefining Airmanship", and his other books on airmanship, provided me with the knowledge and framework I needed to coalesce my thinking about Airmanship 2.0 for personal flyers.

Two diligent editors facilitated the readability of this book: Herbie Ropp and Susan Glancey. They volunteered their valuable time to go through the book with a fine-toothed comb and they offered me their expert opinions on grammar and structure. However, I take full responsibility for any typographical and grammatical errors remaining in the book. Herbie holds an FAA Private Pilot Certificate for single-engine seaplanes only (an unusual qualification), and she is the secretary of the Chicago Flight Instructors Association (CFIA). Susan recently earned her FAA Private Pilot Certificate and she now has over 110 hours in her logbook. In real life, Susan is a language-arts teacher. I owe both of them a great debt of gratitude.

My youngest son, Tyler Koch, designed the cover for this book. Tyler also provided the technical expertise needed to format the book for electronic and print production. He saved me countless hours of frustration and angst. I greatly appreciate his contributions to the project.

Resources

Note: You can type the listed Web addresses into a Web browser to get to the resource, or you can go to:

www.airmanshipexcellence.org/PFG1resources.htm

You will find active links to the resources. Don't forget to bookmark the page so that you can easily return to it.

Preface: *Reader's Preflight Briefing*

Earnest K. Gann: http://en.wikipedia.org/wiki/Ernest_K._Gann

Douglas DC-3s: http://en.wikipedia.org/wiki/Douglas_DC-3

Douglas DC-4s: http://en.wikipedia.org/wiki/Douglas_DC-4

Consolidated C-87 Liberator Express: http://en.wikipedia.org/wiki/Consolidated_C-87_Liberator_Express

Gann's many books: http://www.nytimes.com/1991/12/21/arts/ernest-k-gann-81-an-author-of-many-books-made-into-films.html

Cessna 182 Skylane: http://en.wikipedia.org/wiki/Cessna_182

Chapter 1: *My Pursuit of More Enjoyable Personal Flying*

Stephen Coonts: http://www.coonts.com

Lockheed Constellations:
http://en.wikipedia.org/wiki/Lockheed_Constellation

B-25: http://en.wikipedia.org/wiki/North_American_B-25_Mitchell

Piper Tri-Pacer: http://en.wikipedia.org/wiki/Piper_PA-20_Pacer

Interstate Cadet: http://en.wikipedia.org/wiki/Interstate_Cadet

Aeronca 7AC Champ: http://en.wikipedia.org/wiki/Aeronca_Champion

Cessna 120: http://en.wikipedia.org/wiki/Cessna_140

Cessna 182: http://en.wikipedia.org/wiki/Cessna_182

Frasca: http://www.frasca.com/products/fixedwingpiston.php

Boeing Stearman Model 75: http://en.wikipedia.org/wiki/Boeing-Stearman_Model_75

C-130s: http://en.wikipedia.org/wiki/Lockheed_C-130_Hercules

Bell H-13 Sioux: http://en.wikipedia.org/wiki/Bell_H-13_Sioux

Beechcraft Debonair: http://www.aopa.org/summit/news/2012/121011newsweeps.html

Hughes 300: http://en.wikipedia.org/wiki/Sikorsky_S-300

Boeing 727: http://en.wikipedia.org/wiki/Boeing_727

Beechcraft D-18: http://en.wikipedia.org/wiki/Beechcraft_Model_18

Cessna 177 Cardinal: http://en.wikipedia.org/wiki/Cessna_177_Cardinal

Grumman American AA-1 Yankee Clipper:
http://en.wikipedia.org/wiki/Grumman_American_AA-1

DC-8: http://en.wikipedia.org/wiki/Douglas_DC-8

B-17s: http://en.wikipedia.org/wiki/Boeing_B-17_Flying_Fortress

Bell OH-58 Kiowa: http://en.wikipedia.org/wiki/Bell_OH-58_Kiowa

Bell UH-1 Iroquois: http://en.wikipedia.org/wiki/Bell_UH-1_Iroquois

Seagull: http://www.delta-club-82.com/bible/648-hang-glider-seagull-3.htm

Lake Elsinore: http://www.youtube.com/watch?v=fat_WQl77p8

Skylab: http://en.wikipedia.org/wiki/Skylab

Apollo Command/Service Module:
http://en.wikipedia.org/wiki/Apollo_Command/Service_Module

Chandelle:
http://www.airventuremuseum.org/collection/aircraft/3Chandelle%20Standard%20Hang%20Glider.asp

A-Frame: http://en.wikipedia.org/wiki/Hang_gliding

Hiller OH-13 Raven: http://en.wikipedia.org/wiki/Hiller_OH-23_Raven

Cessna L-19/O-1 Bird Dog: http://en.wikipedia.org/wiki/Cessna_O-1_Bird_Dog

de Havilland DHC-2 Beaver:
http://en.wikipedia.org/wiki/De_Havilland_Canada_DHC-2_Beaver

Sikorsky CH-34 Choctaw: http://en.wikipedia.org/wiki/Sikorsky_H-34

Sky Bob: http://www.airmanshipexcellence.org/skybob.htm

Norge Ski Club: http://www.norgeskiclub.com

Boeing 737-200: http://en.wikipedia.org/wiki/Boeing_737

DC-9s: http://en.wikipedia.org/wiki/McDonnell_Douglas_DC-9

Johnson Space Center:
http://www.nasa.gov/centers/johnson/home/index.html

Airline Pilot Magazine:
http://www.alpa.org/AboutALPA/AirLinePilotMagazine/tabid/2267/Default.aspx

American Institute of Aeronautics and Astronautics: https://www.aiaa.org

Embry-Riddle Aeronautical University's: http://www.erau.edu

Dr. Robert Gilruth: http://en.wikipedia.org/wiki/Robert_R._Gilruth

United Flight 173: http://rbogash.com/Stories/UAL_PDX.html

Space Shuttle Columbia: http://en.wikipedia.org/wiki/STS-1

American Society of Aerospace Pilots:
http://www.csmonitor.com/1982/0803/080303.html

Boeing 747: http://en.wikipedia.org/wiki/Boeing_747

Boeing 737-300: http://en.wikipedia.org/wiki/Boeing_737_Classic

Cessna 337 Skymaster: http://en.wikipedia.org/wiki/Cessna_Skymaster

DC-10: http://en.wikipedia.org/wiki/McDonnell_Douglas_DC-10

AAR Corp: http://en.wikipedia.org/wiki/AAR_Corp

Challenger-600: http://en.wikipedia.org/wiki/Bombardier_Challenger_600

Cessna 421: http://en.wikipedia.org/wiki/Cessna_421

Cessna 402: http://en.wikipedia.org/wiki/Cessna_402

Cessna Citation: http://en.wikipedia.org/wiki/Cessna_Citation_I

T-38: http://en.wikipedia.org/wiki/Northrop_T-38_Talon

Cessna 172: http://en.wikipedia.org/wiki/Cessna_172

Level D simulators: http://en.wikipedia.org/wiki/Full_flight_simulator

Cessna 441 Conquest II: http://en.wikipedia.org/wiki/Cessna_441

The Aerospace Trust: www.aerospacetrust.org

False Security: The Real Story About Airline Safety:
http://www.paperbackswap.com/False-Security-Real-David-Koch-Christine-Koch/book/0972699112/

AvWorld FliteMatrix:
http://www.flightglobal.com/FlightPDFArchive/2004/2004-09%20-%200384.PDF

Eclipse 500: http://en.wikipedia.org/wiki/Eclipse_500

GLASS Simulator Center: http://www.glasssimulator.com

Resources

Beech Bonanza: http://en.wikipedia.org/wiki/Beechcraft_Bonanza

Frasca 142 TruFlite: http://frasca.com/products/truflite.php

2009 Nall Report: http://www.aopa.org/-/media/Files/AOPA/Home/News/All%20News/2011/January/Safety%20Pilot/09nall.pdf

Air Safety Institute: http://www.aopa.org/asf/

National Transportation Safety Board: http://www.ntsb.gov/investigations/reports_aviation.html

FAA Wings Program: www.faasafety.gov

David Greenberg: http://airmanshipexcellence.org/Board%20of%20Advisors.htm

T-37: http://en.wikipedia.org/wiki/Cessna_T-37_Tweet

Korean Airlines: http://www.nytimes.com/2002/03/26/business/new-standards-mean-korean-air-is-coming-off-many-shun-lists.html

Cargo360: http://en.wikipedia.org/wiki/Cargo_360

***Redefining Airmanship*:** http://www.amazon.com/Redefining-Airmanship-Anthony-Kern/dp/0070342849/ref=sr_1_wsc1?ie=UTF8&qid=1355159348&sr=8-1-wordsplitter&keywords=%C2%93Redefining+Airmanship%C2%94

Tony Kern: http://www.linkedin.com/pub/tony-kern/10/1b8/1a6

The Kern Airmanship Model: http://airmanshipexcellence.org/Flying%202.0/Kernmodel.htm

Mr. David Shadle: http://airmanshipexcellence.org/Board%20of%20Advisors.htm

Falcon 20s: http://en.wikipedia.org/wiki/Dassault_Falcon_20

Elliott Aviation: http://www.elliottaviation.com

King Air 200: http://en.wikipedia.org/wiki/Beechcraft_Super_King_Air

Falcon 10s: http://www.dassault-aviation.com/en/passion/aircraft/civil-dassault-aircraft/mystere-falcon-10-100.html?L=1

King Air 300/350: http://en.wikipedia.org/wiki/Beechcraft_Super_King_Air

Beech 1900 Airliner: http://en.wikipedia.org/wiki/Beechcraft_1900

FAA Safety Team: www.faasafety.gov

Capt. Arnold Quast:
http://airmanshipexcellence.org/Board%20of%20Advisors.htm

Capt. Bill Brand:
http://airmanshipexcellence.org/Training%20Operations%20Committee%20(TOC).htm

Capt. Richard Sternal:
http://airmanshipexcellence.org/Safety%20and%20Standards%20Committee%20(SSC).htm

Mr. John Keiper:
http://airmanshipexcellence.org/Flight%20Operations%20Support%20Committee.htm

CFAE Website: http://airmanshipexcellence.org

Capt. Dave's Hangar: http://captaindaveshangar.wordpress.com

Redefining Airmanship Wings Program Webinar Series:
http://airmanshipexcellence.org/Archive.htm

Airmanship Archives: http://airmanshipexcellence.org/Archive.htm

Kishwaukee College: www.kishwaukeecollege.edu/

National Association of High School Aviation Clubs: www.nahsac.org

Lewis University: http://www.lewisu.edu

Southern Illinois University: http://siu.edu

Women In Aviation International: http://www.wai.org/welcome.cfm

Aviation and Space Education Program:
http://www.faa.gov/education/contact/

General Aviation News: http://www.generalaviationnews.com/2011/08/an-aviation-club-in-every-high-school/

Dr. Frank Bacon:
http://airmanshipexcellence.org/Board%20of%20Advisors.htm

Cessna 170: http://en.wikipedia.org/wiki/Cessna_170

Piper Archer: http://en.wikipedia.org/wiki/Piper_PA-28_Cherokee

Planned Innovation Institute: http://plannedinnovationinstitute.com

Planned Innovation: http://plannedinnovationinstitute.com/mission.html

Chicago Flight Instructors Association (CFIA):
http://www.cfia.us/history.htm

EAA International Aerobatic Club (IAC):
http://en.wikipedia.org/wiki/International_Aerobatic_Club

EAA Spirit of St. Louis:
http://www.airventuremuseum.org/collection/aircraft/4Ryan%20Spirit%20of%20St%20Louis%20Replica.asp

Diamond DA-20s: http://en.wikipedia.org/wiki/Diamond_DA20

Kern Airmanship Model In Detail:
http://airmanshipexcellence.org/Archive.htm

Airmanship 2.0: http://airmanshipexcellence.org/Archive.htm

Airmanship 2.0 Case Studies: http://airmanshipexcellence.org/Archive.htm

Airmanship Challenge Program:
http://airmanshipexcellence.org/Archive.htm

John Nowicki:
http://airmanshipexcellence.org/Aircraft%20Selection%20and%20Maintenance%20Committee%20(ASMC).htm

Tim Perry: http://www.linkedin.com/in/timothyfperry

Beech Turbo Mentor T-34Cs: http://en.wikipedia.org/wiki/Beechcraft_T-34_Mentor

Beech King Air T-44As: https://en.wikipedia.org/wiki/Beechcraft_King_Air

P-3 Orion: https://en.wikipedia.org/wiki/Lockheed_P-3_Orion

Gulfstream C20Gs: http://en.wikipedia.org/wiki/Gulfstream_IV

Airbus A320/319: http://en.wikipedia.org/wiki/Airbus_A320_family

Airmanship Development Support Organization Demonstrator:
http://airmanshipexcellence.org/Flying%202.0/ADSOD.htm

Peter Halasz: http://www.phyxius.com/default.htm

Michael Lockett: http://www.michaelmlockett.com/index.php

CPAaviation.com: http://cpaaviation.com

Advocates For Aviation Safety Foundation: http://www.afasf.org

Andrew Roccasalva:
http://airmanshipexcellence.org/Airman%20Outreach%20Committee.htm

Joshua Allison:
http://airmanshipexcellence.org/Airman%20Outreach%20Committee.htm

Chapter 2: *Airmanship 2.0*

Richard Bach: http://en.wikipedia.org/wiki/Richard_Bach

Center For Airmanship Excellence: http://airmanshipexcellence.org

Redefining Airmanship: http://www.amazon.com/Redefining-Airmanship-Anthony-Kern/dp/0070342849/ref=sr_1_wsc1?ie=UTF8&qid=1355159348&sr=8-1-wordsplitter&keywords=%C2%93Redefining+Airmanship%C2%94

Kern Airmanship Model:
http://airmanshipexcellence.org/Flying%202.0/Kernmodel.htm

Kern Airmanship Model In Detail Wings Program Webinar:
http://airmanshipexcellence.org/Flying%202.0/Kernmodel.htm

Airmanship Development Support Organization:
http://airmanshipexcellence.org/Flying%202.0/ADSO.htm

Airmanship 1.0: http://airmanshipexcellence.org/Flying%202.0/Airmanship1.htm

Daedalus and Icarus: http://en.wikipedia.org/wiki/Daedalus

Flight Risk Assessment Tool:
http://www.faa.gov/other_visit/aviation_industry/airline_operators/airline_safety/info/all_infos/media/2007/info07015.pdf

Douglas DC-8-61: http://en.wikipedia.org/wiki/Douglas_DC-8

FAA TV: http://www.faa.gov/tv/?mediaid=447

FAA: http://faasafety.gov

Aircraft Owners and Pilots Association: http://www.aopa.org

Experimental Aircraft Association:
https://secure.eaa.org/apps/joinrenew/join.aspx?sc=219

Gen. Charles "Chuck" Yeager: http://www.chuckyeager.com

Bell X-1: http://en.wikipedia.org/wiki/Bell_X-1

Airmanship Development Support Organization:
http://airmanshipexcellence.org/Flying%202.0/ADSO.htm

Personal Airmanship Development Plan:
http://airmanshipexcellence.org/Flying%202.0/Airmanshipdevelopmentplan.htm

Airmanship Challenges:
http://airmanshipexcellence.org/Airmanship%20Challenges/Airmanship%20Challenge.htm

Kern Airmanship Model In Detail Wings Program Webinar Archive:
http://airmanshipexcellence.org/Archive.htm

crosswind component: http://en.wikipedia.org/wiki/Crosswind

Federal Aviation Regulation 61.57:
http://rgl.faa.gov/Regulatory_and_Guidance_Library/rgFar.nsf/FARSBySectLookup/61.57

Command/Leadership/Resource Management :
http://www.avops.com/CRMCombined.pdf

Safety Management System: http://www.faa.gov/about/initiatives/sms/

Chapter 3: *Airmanship 2.0 In Practice*

Charles Lindbergh: https://en.wikipedia.org/wiki/Charles_Lindbergh

NAHSAC.org: www.nahsac.org

simulators: http://www.frasca.com/products/cirrussr20.php

computer-training carrels: http://www.novadesk.com/blog/bid/54790/The-Study-Carrel-Optimizing-This-Important-Nook-For-Independent-Work

Buck Rogers: http://en.wikipedia.org/wiki/Buck_Rogers

Gulfstream G650: http://www.gulfstream.com/products/g650/

Cirrus SR22: http://cirrusaircraft.com/sr22/

Garmin G1000 Perspective:
http://www.whycirrus.com/advancements/Perspective%20comparison%20to%20off-the-shelf%20G1000.pdf

Vertical Power VP-400: http://verticalpower.com/vp-400/

Cirrus Aircraft Parachute System: http://www.youtube.com/watch?v=i_B--xSUxBA

Astronics Max-Viz 600: http://www.youtube.com/watch?v=9bkYXdlLYJ8

Endoscope:
http://www.aopa.org/aopalive/?watch=A2eTZoMjqZKhVK3AwLF4K8HiXahzBzqN

Cirrus SR20: http://cirrusaircraft.com/sr20/

Flight Design CTLSi: http://flightdesignusa.com/aircraft/ctls/

Garmin G3X: https://buy.garmin.com/shop/shop.do?pID=63892

Garmin GTN 750: https://buy.garmin.com/shop/shop.do?pID=67886

Spidertracks Spider S3: http://www.spidertracks.com/store/spiders/spider-s3

Angle-Of-Attack Indicator:
http://www.youtube.com/watch?v=DCerL8ljRwk

Head-Up Displays: http://www.patavionics.com/index.php/en/

Resources

Airbags: http://www.amsafe.com/products-services/general-aviation/

Diamond DA-42 Twinstar: http://en.wikipedia.org/wiki/Diamond_DA42

AirCam: http://www.aircam.com

Amphibious floats: http://www.youtube.com/watch?v=3KN2LOgN9kU

Waco YMF-5D Super: http://www.wacoaircraft.com/ymf/

Piper J3 Cub: http://en.wikipedia.org/wiki/Piper_J-3_Cub

CubCrafters Sport Cub S2: http://www.cubcrafters.com/sportcubs2

Desktop Simulator: http://www.youtube.com/watch?v=DmS76txgsW0

Mercedes-Benz Sprinter: http://www.mbsprinterusa.com/sprinter/passenger-van

T-hangars: http://en.wikipedia.org/wiki/Tee_hangar

Beech King Air B90: http://en.wikipedia.org/wiki/Beechcraft_King_Air

Beech Baron: http://en.wikipedia.org/wiki/Beechcraft_Baron

Beech Premier 1A: http://en.wikipedia.org/wiki/Beechcraft_Premier_I

Cub simulator: http://www.youtube.com/watch?v=BVnlfra7gk0

Taildragger: http://en.wikipedia.org/wiki/Conventional_landing_gear

Transponder: http://en.wikipedia.org/wiki/Transponder_(aviation)

Appareo Vision 1000:
http://www.appareo.com/primarymenu/products/alerts-flight-data-monitoring/appareo-vision-1000/

Appareo's Aircraft Logging and Event Recording for Training and Safety: http://www.appareo.com/cat/primarymenu/products/alerts-flight-data-monitoring/

T-6: http://en.wikipedia.org/wiki/North_American_T-6_Texan

P-51: http://en.wikipedia.org/wiki/North_American_P-51_Mustang

B-25: http://en.wikipedia.org/wiki/North_American_B-25_Mitchell

B-17: http://en.wikipedia.org/wiki/Boeing_B-17_Flying_Fortress

Tolhuin: http://en.wikipedia.org/wiki/Tolhuin

Young Eagles Program: http://www.youngeagles.org

Volunteer Pilots Association: http://www.volunteerpilots.org

LightHawk: http://www.lighthawk.org

FAA Wings Pilot Proficiency Program:
https://www.faasafety.gov/WINGS/pppinfo/default.aspx

Aircraft Owners and Pilots Association: http://www.aopa.org

The Flight Training Experience:
http://download.aopa.org/epilot/2011/AOPA_Research-The_Flight_Training_Experience.pdf

Flying Club Network: http://www.aopa.org/Media-Relations/Press-Releases/2012/AOPA-UNVEILS-DETAILS-ON-NEW-FLYING-CLUB-NETWORK

AOPA Center To Advance The Pilot Community:
http://www.aopa.org/CAPComm/

Embraer Phenom 100: http://en.wikipedia.org/wiki/Embraer_Phenom_100

Bombardier Learjet 85: http://en.wikipedia.org/wiki/Learjet_85

Chapter 4: *Your Pursuit of More Enjoyable Personal Flying*

Amelia Earhart: http://www.ameliaearhart.com

via email: dkoch@airmanshipexcellence.org

ADSOD Pilot Participant Application:
http://airmanshipexcellence.org/Forms/ADSODapplication.html

*Thank you for taking your valuable time to read this book.
Now, I suggest that you take the time to think about the ideas and concepts in
it and to discuss the pursuit of airmanship excellence
with your friends who also have a passion for personal flying.*

*You can find a plethora of additional information about the pursuit of
airmanship excellence on the Center For Airmanship Excellence Website
(www.airmanshipexcellence.org).*

*If you have any questions, concerns, suggestions or comments,
please feel free to pass them along to me at dkoch@airmanshipexcellence.org.*

*When you're ready to start down the track to becoming an excellent airman or
airwoman, let me know if I can help you to find
the track that's right for you.*

Until then, please remember, if you're going to fly, do it right.

About The Author

Capt. David C. Koch has been a personal flyer since 1958 and he has been an aerospace-industry professional for over 48 years. He has extensive experience as an aviation manager and consultant. He has many aviation-related entrepreneurial ventures to his credit. He has written numerous aerospace-industry-related articles, papers and technical documents including business plans and business cases. He is the author of "A Personal Flyer's Guide to More Enjoyable Flying" and the co-author of the book "False Security: The Real Story About Airline Safety". He has many media appearances to his credit and he has been a featured speaker at various aerospace-industry events and meetings. He is also the creator, writer, producer and host of several aviation-related video productions.

He has served as an airline (United Airlines), military (U.S. Army) and corporate pilot and has accumulated over 18,000 flight hours. Capt. Koch has been an FAA Certified Flight Instructor since 1965 and he served as an FAA-designated Training Center Evaluator for the world's largest corporate-pilot training vendor. He currently enjoys personal flying in many types of private aircraft.

Capt. Koch has served in the following aerospace-industry-related capacities: Special Assistant to the president of United Airlines for Space Shuttle acquisition, Founding National Chairman of the American Society of Aero Space Pilots (ASAP), Chairman of the Publicity Committee of the Houston Chapter of the American Institute of Aeronautics and Astronautics (AIAA), Founding President of the Houston Chapter of the Aviation/Space Writers Association (AWA), Founding Chairman of the Air Line Pilots Association Professional Outlook Committee (ALPA—United Airlines), Founding Chairman of the United Pilots Buyout Committee, Founding Chairman of Houston Space Week, CEO of All-States Aviation, CEO of Apollo Skysails, CEO of AeroComm, CEO of Aerospace International Companies, CEO of AvWorld Public Charters, CEO of Cloud 9 Air Travelers Club, CEO of AvWorld FliteMatrix and he is currently the Managing Trustee of The Aerospace Trust. Capt. Koch serves on the Kishwaukee College Aviation Advisory Committee and as an FAA Safety Team representative. He is also a member of the FAA Wings Industry Advisory Committee. Capt. Koch currently serves as the Executive Director of the Center For Airmanship Excellence.

www.ingramcontent.com/pod-product-compliance
Lightning Source LLC
Chambersburg PA
CBHW071305110426
42743CB00042B/1176